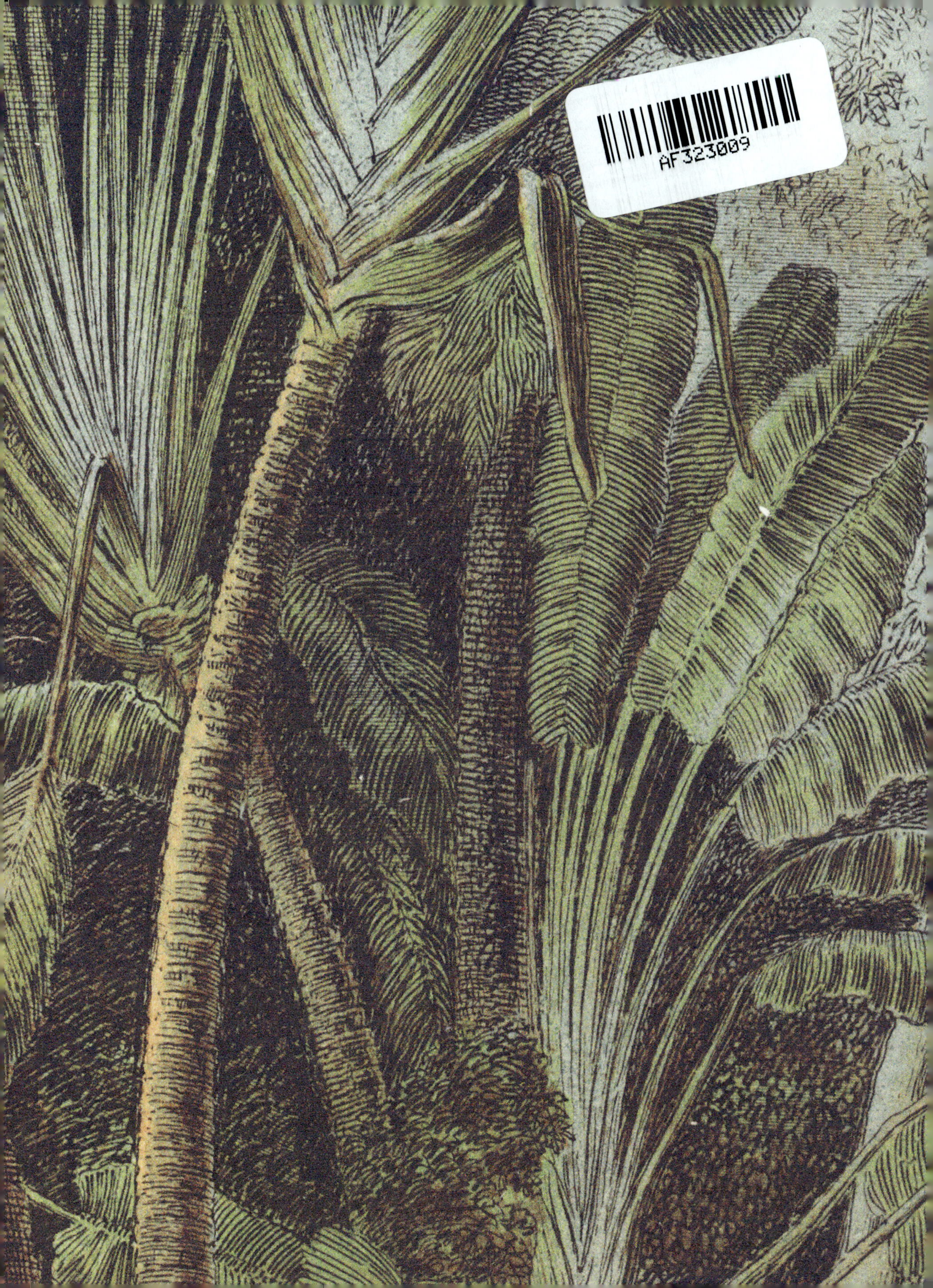

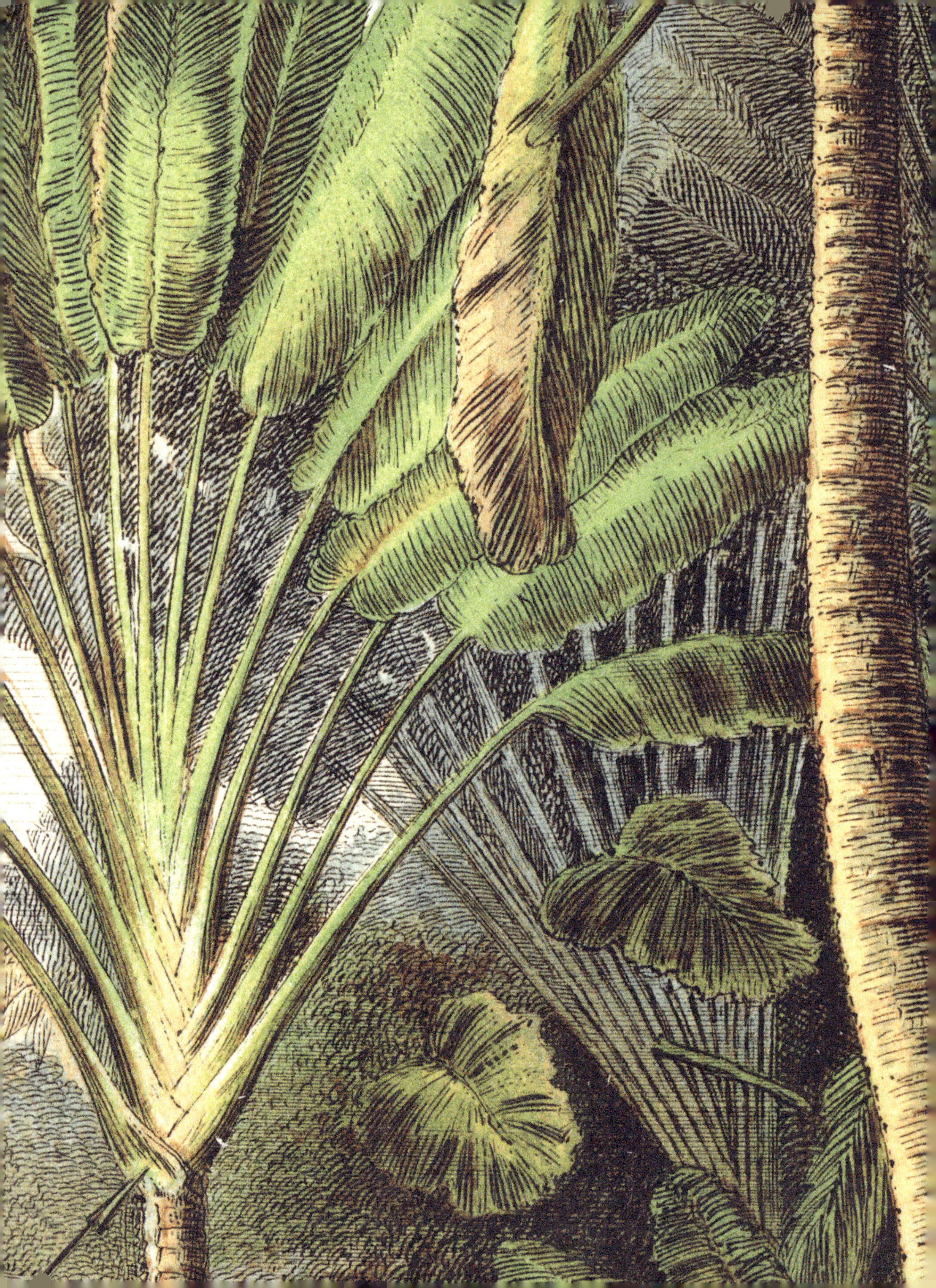

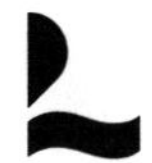

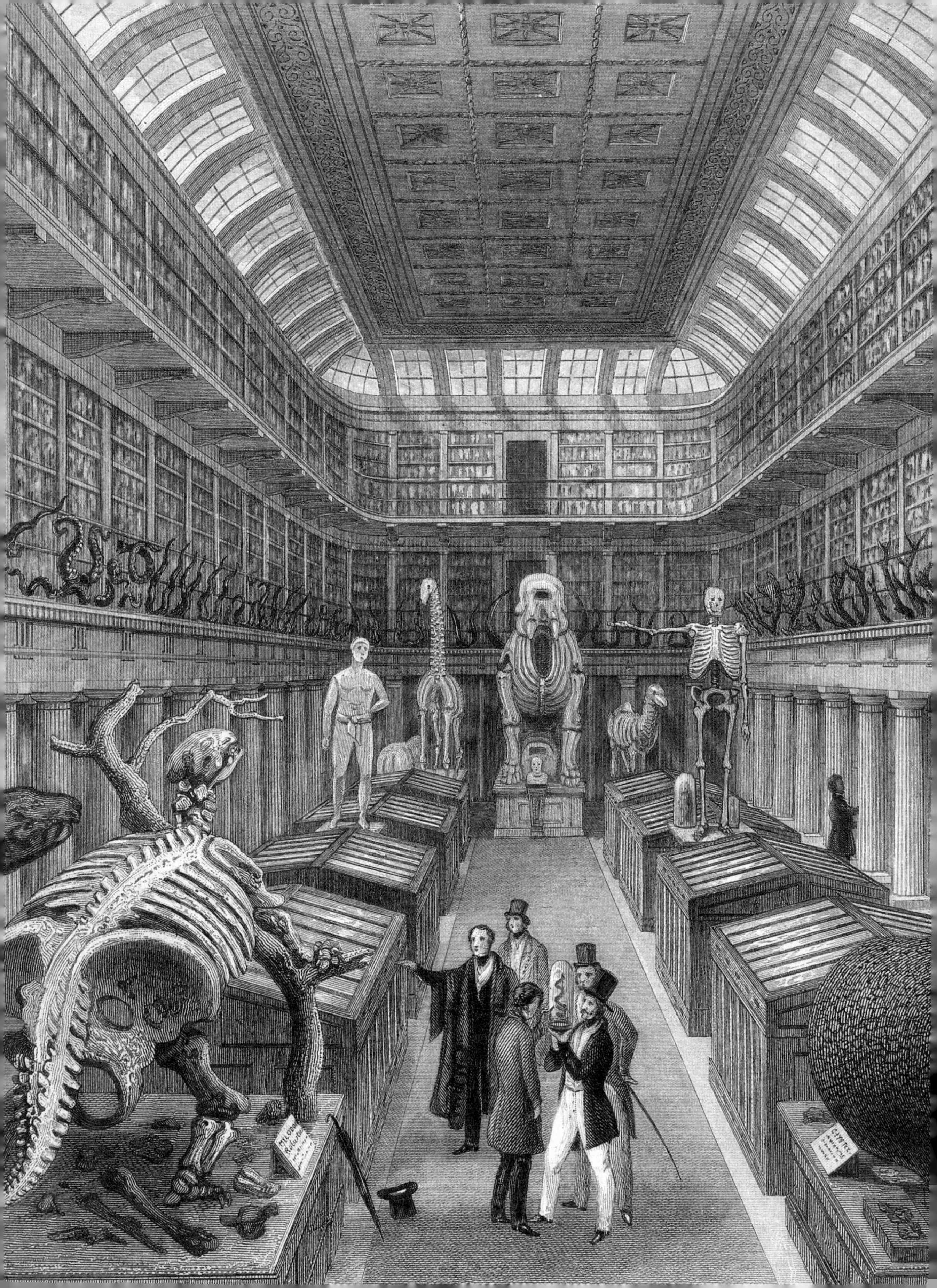

DINOSAURS ARE COLLECTIBLE

DIGGING FOR DINOSAURS: THE ART, THE SCIENCE

THIJS DEMEULEMEESTER / KOEN STEIN

Lannoo

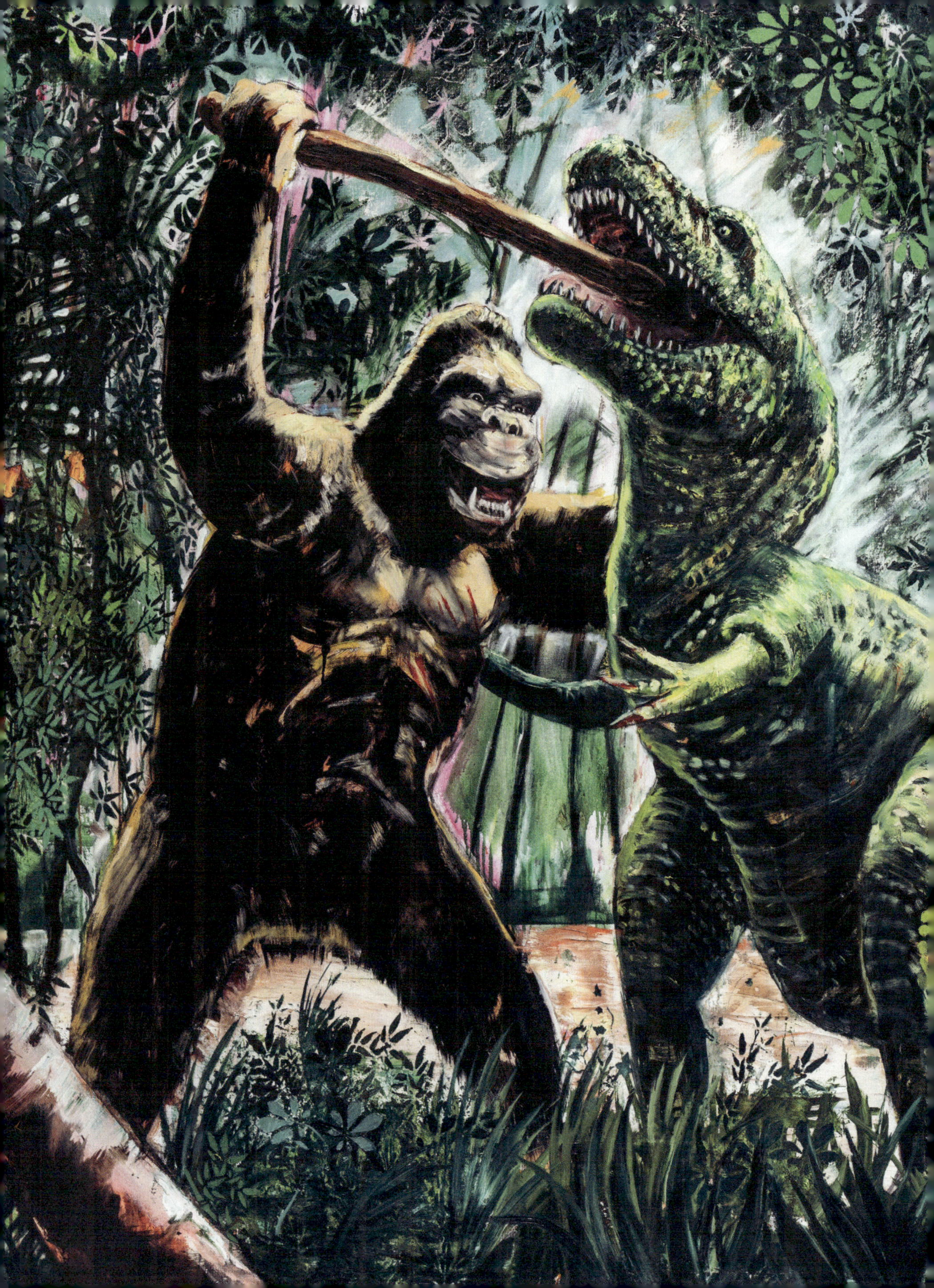

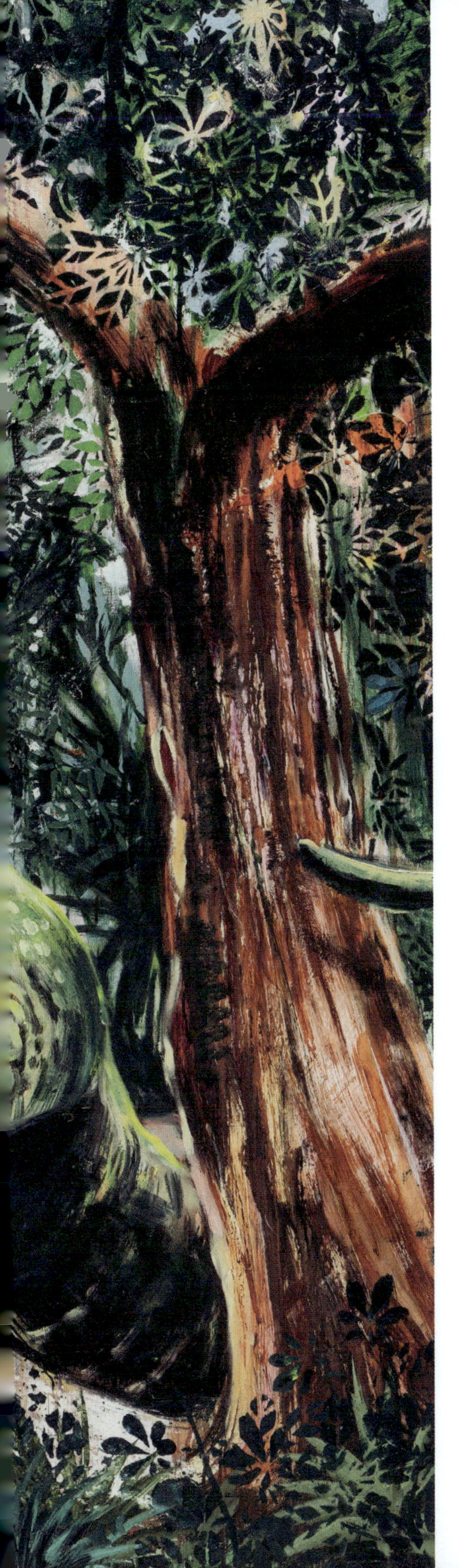

CONTENTS

<
Tom Liekens, *The Land That Time Forgot*, 2018.

>
A *Mosasaurus* skull in the salon of Italian
filmmaker Francesco Invernizzi.

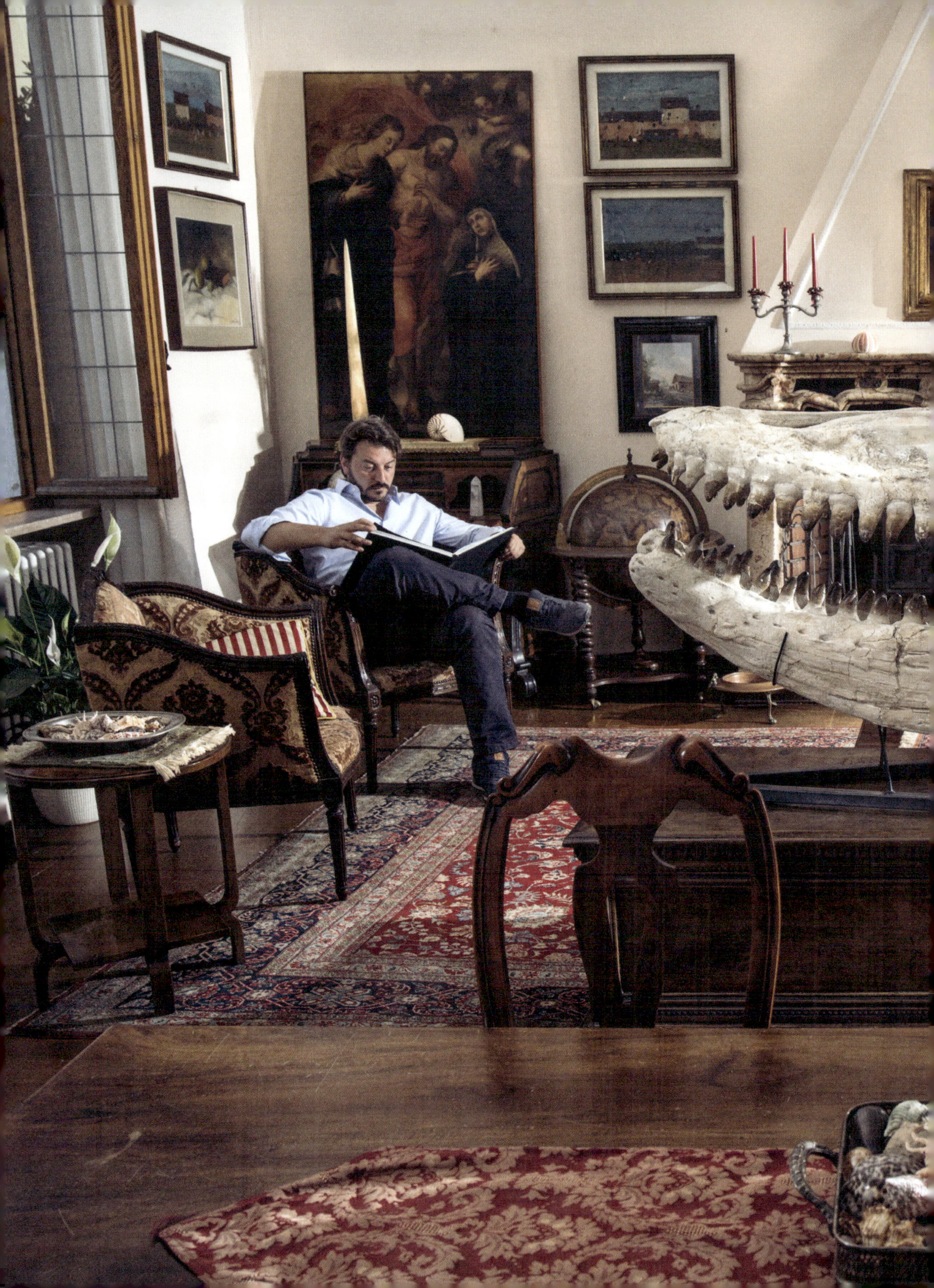

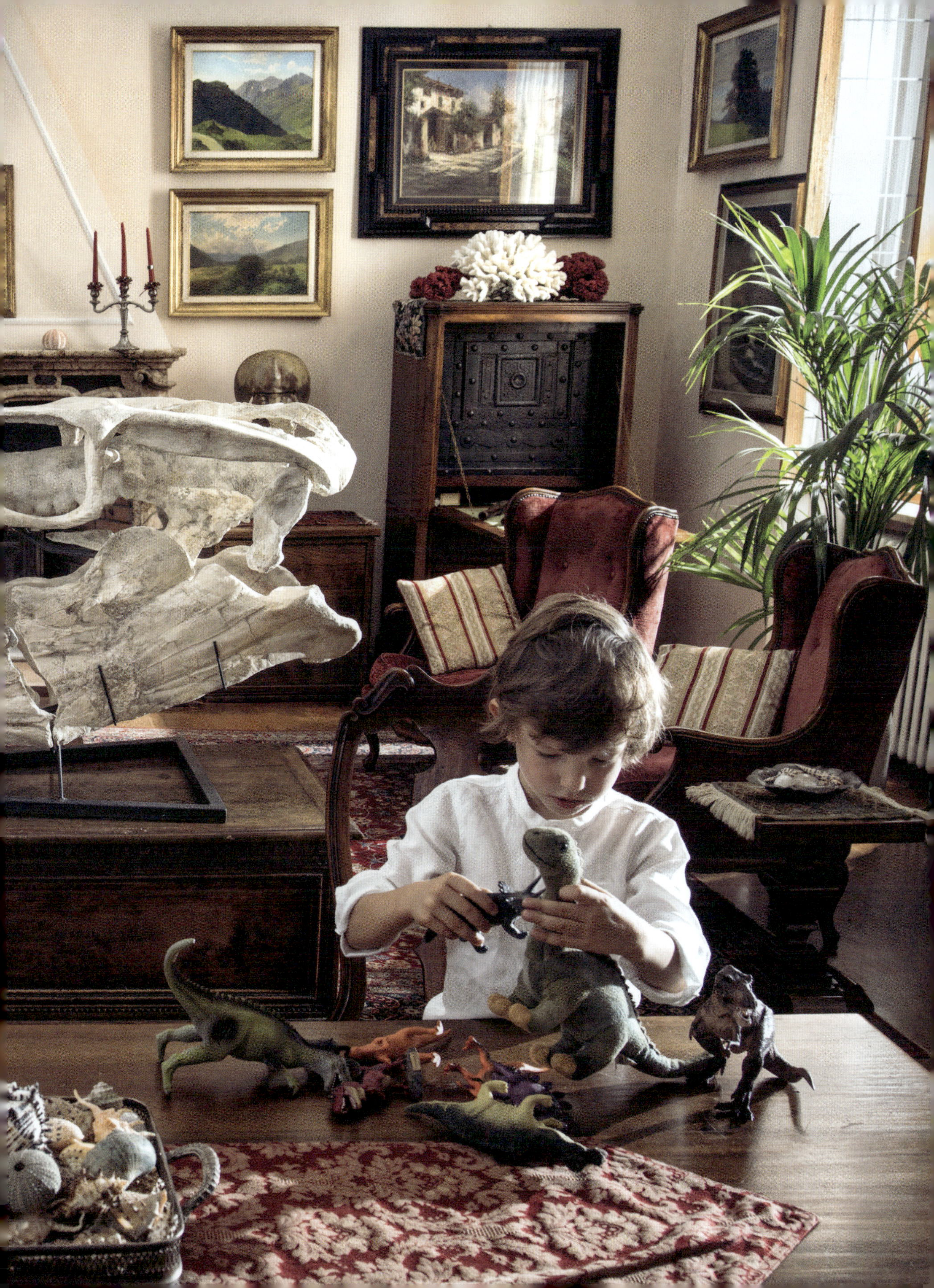

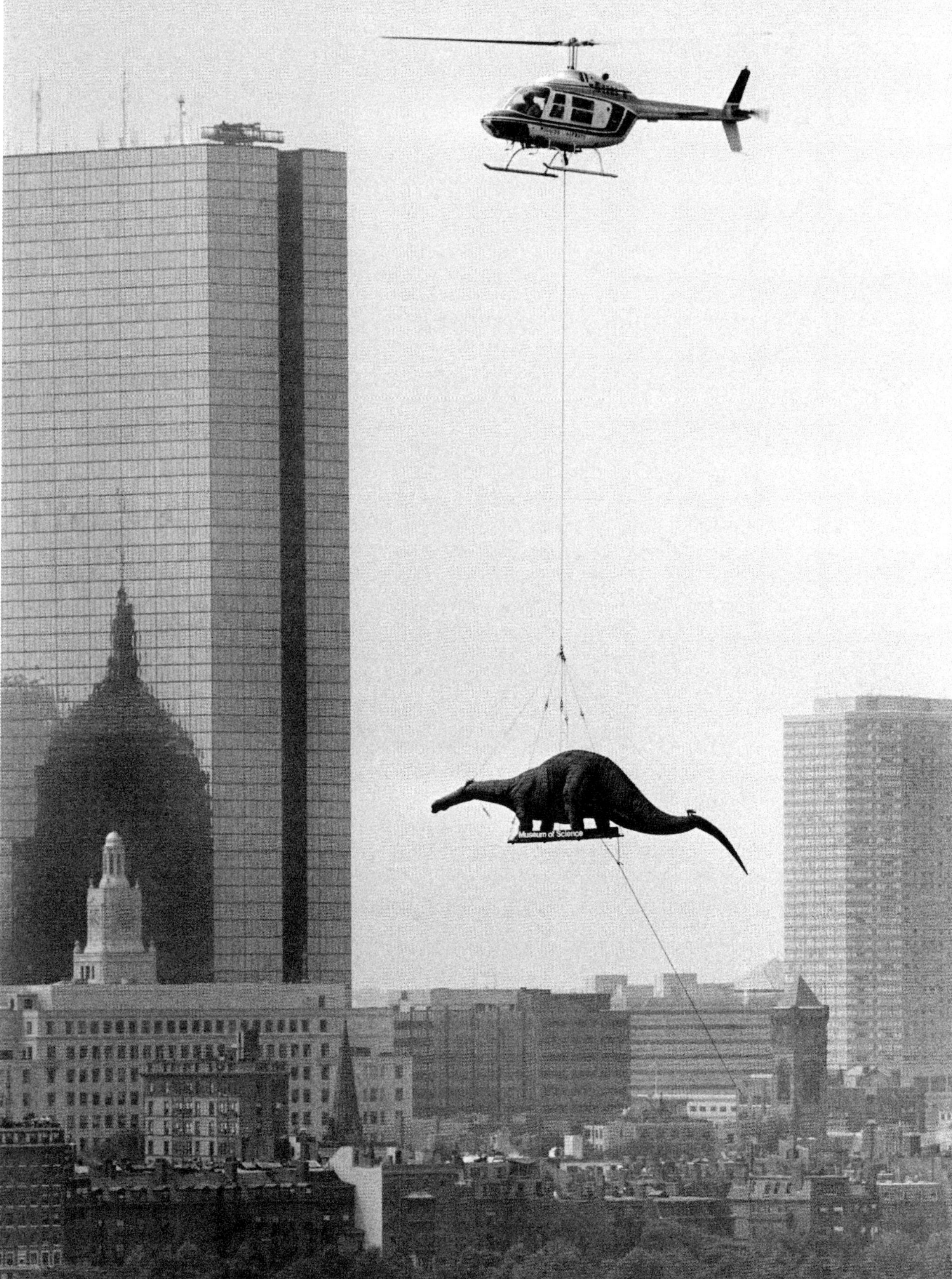

Museum of Science

COLLECTING DINOSAURS

BUYING AND SELLING DINOSAURS

—

'Dinosaurs are not tied to Eastern or Western culture.

The continents as we know them today did not even

exist when they roamed the Earth.

That is why dinosaurs are like the Moon:

they belong to everyone.'

Iacopo Briano, natural history specialist, 2021

—

Let's begin with a true story. One day, a parcel delivery guy turns up on our doorstep with a large box in his hands. We open it, only to discover that it contains a turd. Our first thought? This must be someone's idea of a joke. But on closer inspection, it turns out to be a belated birthday gift. The droppings are hard and odourless. The accompanying greetings card says it's a fossil more than 70 million years old, and that it comes from an (unspecified) dinosaur. We didn't flush the coprolite – for that is what fossilised dinosaur excrement is called – down the toilet. It stands 'proudly' in our display cabinet. And every time children, teens or adults spot it, they react in the same way: first horror, then disbelief and, finally, fascination.

<

A helicopter delivers a dinosaur to the Boston Museum of Science in 1984.

Dinosaur fossils make us smile. Well, actually, anything 'dinosaur' does. Particularly when you realise you can collect them, even if it's only bits and pieces. A *Tyrannosaurus rex* tooth costs between €3,000 and €20,000, a *Triceratops* horn around €7,000. A chunk of fossilised poop costs around €50. If you dream of collecting larger dinosaur fossils, you'll need a heftier budget. A jawbone from an *Allosaurus* – the most dangerous predator of the Jurassic Period – can easily cost €50,000. And a well-preserved dinosaur skull can cost between €25,000 and €5 million, depending on its rarity and completeness. Complete skeletons are often sold privately between collectors, but they also sometimes come up for sale at auctions, art fairs or specialised dealers. When they do, they can often count on a lot of interest, from young and old.

'BIG JOHN'

Dinosaur fans lined up around the block to see 'Big John', the *Triceratops horridus* that was sold in October 2021 at the Paris auction house Binoche et Giquello. Remarkably, not only children but older people flocked to see the horned dinosaur in real life. And what a creature it is – measuring 7.15m from head to tail, his skull alone 2.6m long and 2m wide. More remarkable still, the entire skeleton is 60 per cent complete. Impressive statistics for a *Triceratops*, specimens of which rarely appear on the market. The auction house went all out to create a buzz around the sale, selling Big John merchandise, such as T-shirts and face masks, and with a sale catalogue compiled with a flair for the dramatic. 'There's no question that the most life-threatening situation for a carnivore would be a run-in with a *Triceratops*. It's probably the most lethal creature to have walked the Earth,' the famous American palaeontologist (a scientist who studies fossils) Robert T. Bakker is quoted as saying in the catalogue. The hype worked, because the prehistoric colossus sold for €6.6 million to a telephone bidder, an anonymous, private collector. Quite a sizeable sum for a fossil that has little scientific importance in the history of palaeontology. Geologist Walter W. Stein, leader of the excavations at Mud Butte Ranch in South Dakota, did not discover the creature until 2014. And Big John was certainly not the first *Triceratops* to be unearthed in the Hell Creek Formation: dinosaur remains have been excavated there since the end of the 19th century. Palaeontologist John Scannella once said: 'It is hard to walk out into the Hell Creek Formation and not stumble upon a triceratops weathering out of a hillside.'

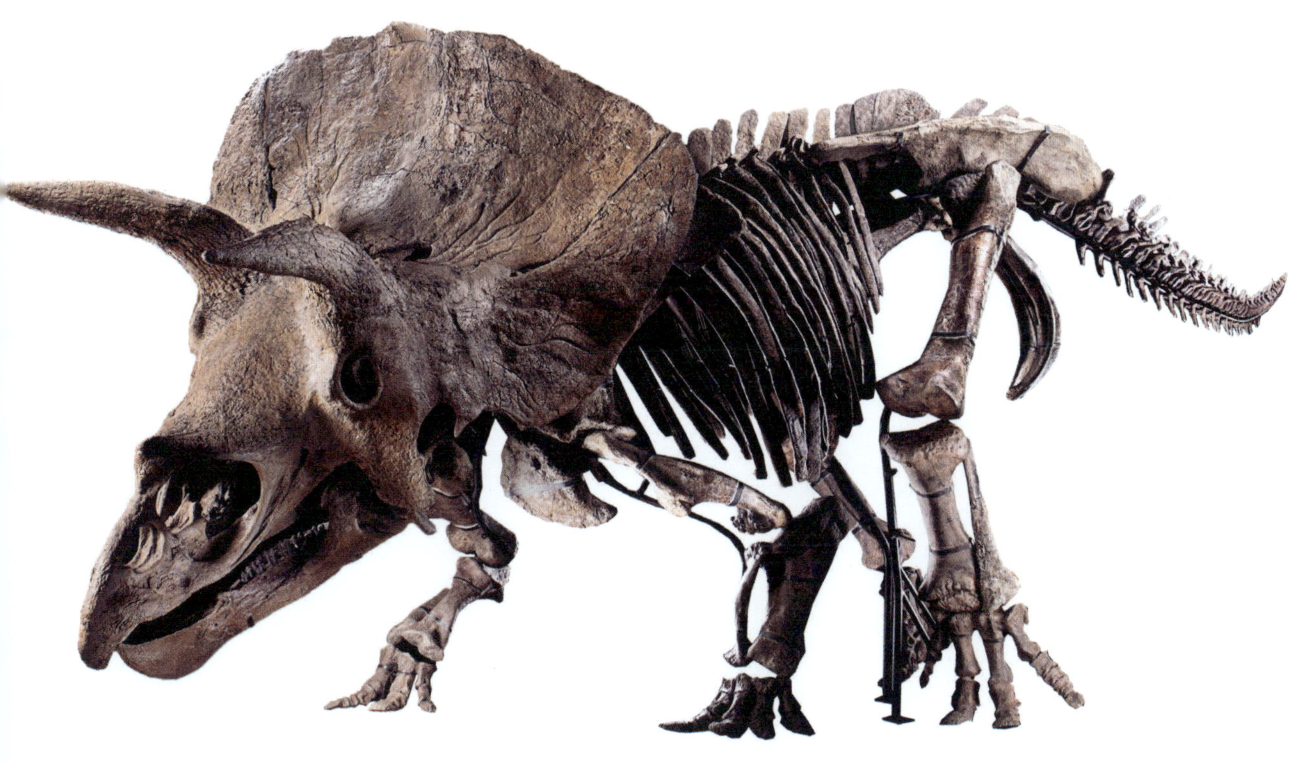

The *Triceratops* 'Big John' was sold for 6.6 million euros in 2021.

Scottish American industrialist and philanthropist Andrew Carnegie donated a cast of *Diplodocus carnegii* to the Musée National d'Histoire Naturelle in Paris in 1908 (see p. 58), of which a cast also is kept at the Naturhistorisches Staatsmuseum in Vienna.

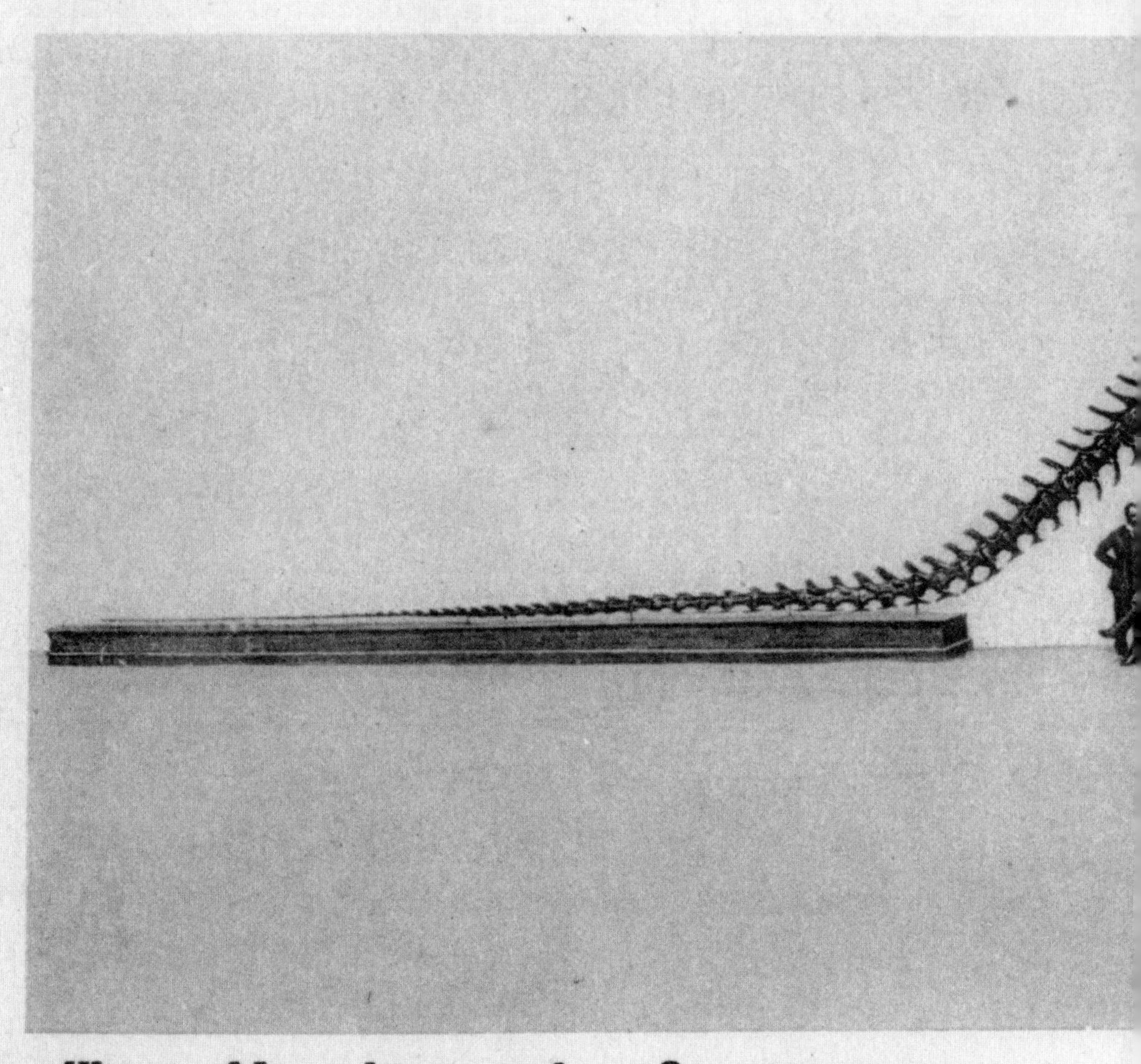

Wien, Naturhistorisches Staatsmuseum

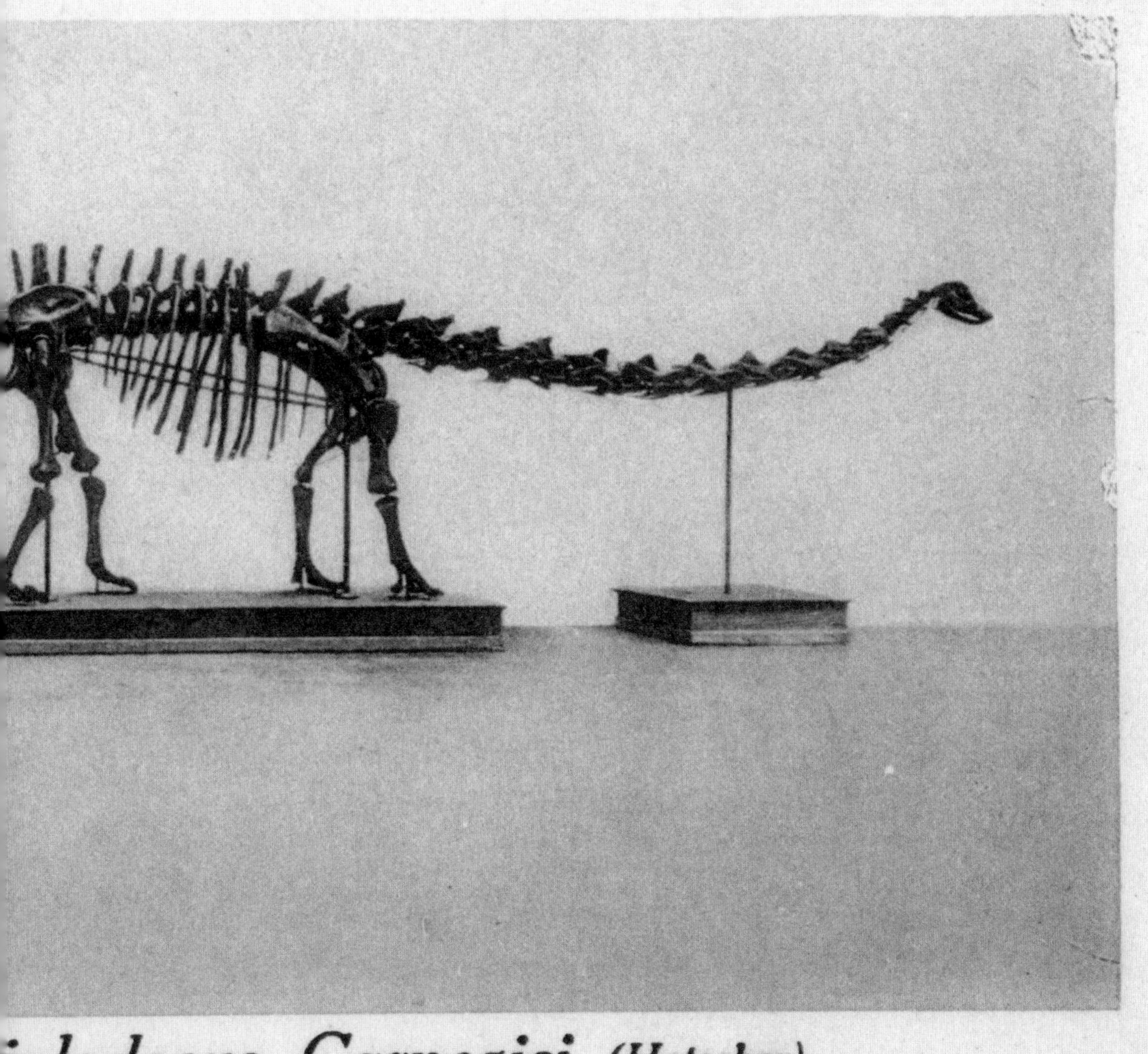

iplodocus Carnegici (Hatscher).
34 m lang.

A year earlier, on 13 October 2020, a complete *Allosaurus* that went under the hammer at Binoche et Giquello also garnered a huge amount of interest. This 10 by 3.5 million skeleton was valued at €1 million, but ultimately fetched just over €3 million. Of all the dinosaur species, allosaurs are among the more common, and come to light relatively often. For the public at large, the *Allosaurus* lacks the wall power of a *Tyrannosaurus rex*, although that's a little unfair. Together with *Ceratosaurus*, it was one of the most dangerous predators of the Jurassic Period. With teeth as sharp as steak knives, it was at the top of the food chain and built to kill. Walking on its hind legs, it probably attacked its prey with its mouth wide open. This is precisely how the skeleton at the auction was presented, mounted in a dynamic combat stance, primed to attack, unlike old specimens in natural history museums which are often displayed in the 'rigid', 19th-century style. Auction houses selling dinosaur skeletons cleverly capitalise on dinosaurs' cinematic presence, mounting them as though they were film stills from *Jurassic Park*.

According to the natural history experts associated with the auction, 'The *Allosaurus* must have fought ferociously 150 million years ago, because in a fight with another dinosaur, it broke three ribs, which fused together.' Although there's no way of telling whether that epic battle actually took place, it's a story that fires our imaginations. Which is precisely why auction houses have begun to sell dinosaurs, and why today's collectors pay such immense sums in what is a relatively new market. Iacopo Briano, a natural history specialist who works with auction houses such as Binoche et Giquello, understands this universal interest all too well. 'Dinosaurs are not tied to Eastern or Western culture. The continents as we know them today did not even exist when they roamed the Earth. That is why dinosaurs are like the Moon: they belong to everyone.'

A $31 MILLION DINOSAUR

There's no question that the most talked-about dinosaur that sold recently was 'Stan', one of the largest, best-known and most intact skeletons of a *Tyrannosaurus rex* ever discovered. Stan comes with a spectacular story. Scientists discovered he had a broken neck and a hole in his skull, possibly from a bite from a fellow *T. rex*. Stan Sacrison, an amateur palaeontologist who freelanced for the Black Hills Geological Institute, unearthed the skeleton in 1987. In 1992, the year before the release of *Jurassic Park*, excavations began under the direction of Peter Larson.

Stan's 2020 sale at Christie's New York involved a superabundance of stunt work. The dinosaur wasn't on sale at a 'traditional' auction packed with fossils or other natural history collectibles. Christie's pulled out all the stops, crowning Stan the king of their evening sale of modern and contemporary art. The dinosaur rubbed shoulders with sought-after artists such as Jackson Pollock and Paul Cézanne. When Christie's offer a *T. rex* skeleton and a Jackson Pollock at the same evening sale on the same night, there can be no doubt – dinosaur fossils are the new collectibles. The question is, can you put Stan on the same level as a Picasso or a Matisse? Is he equally rare? And will he attract the same wealthy collectors as blue-chip art? Every time, the answer has been a resounding 'yes'. That night, the skeleton sold for the all-time world record of $31 million. Evidently, Christie's cross-disciplinary sales strategy was an electrifying success; overnight, dinosaurs had become as desirable as the works of the modern masters. And that pushed prices up, too, especially for a masterpiece like Stan, one of the world's most lauded dinosaur skeletons. 'Coming face to face with Stan for the first time was awe-inspiring. He was bigger and scarier than I'd imagined,' said James Hyslop, head of the natural history department at Christie's. 'When I first saw him, he was just a jumble of bones in a crate. But over the next eight to 12 hours, I got to see him being assembled, bone by bone. It was breathtaking. It's not often you get so close to one of the most complete *T. rex* skeletons ever found.'

———

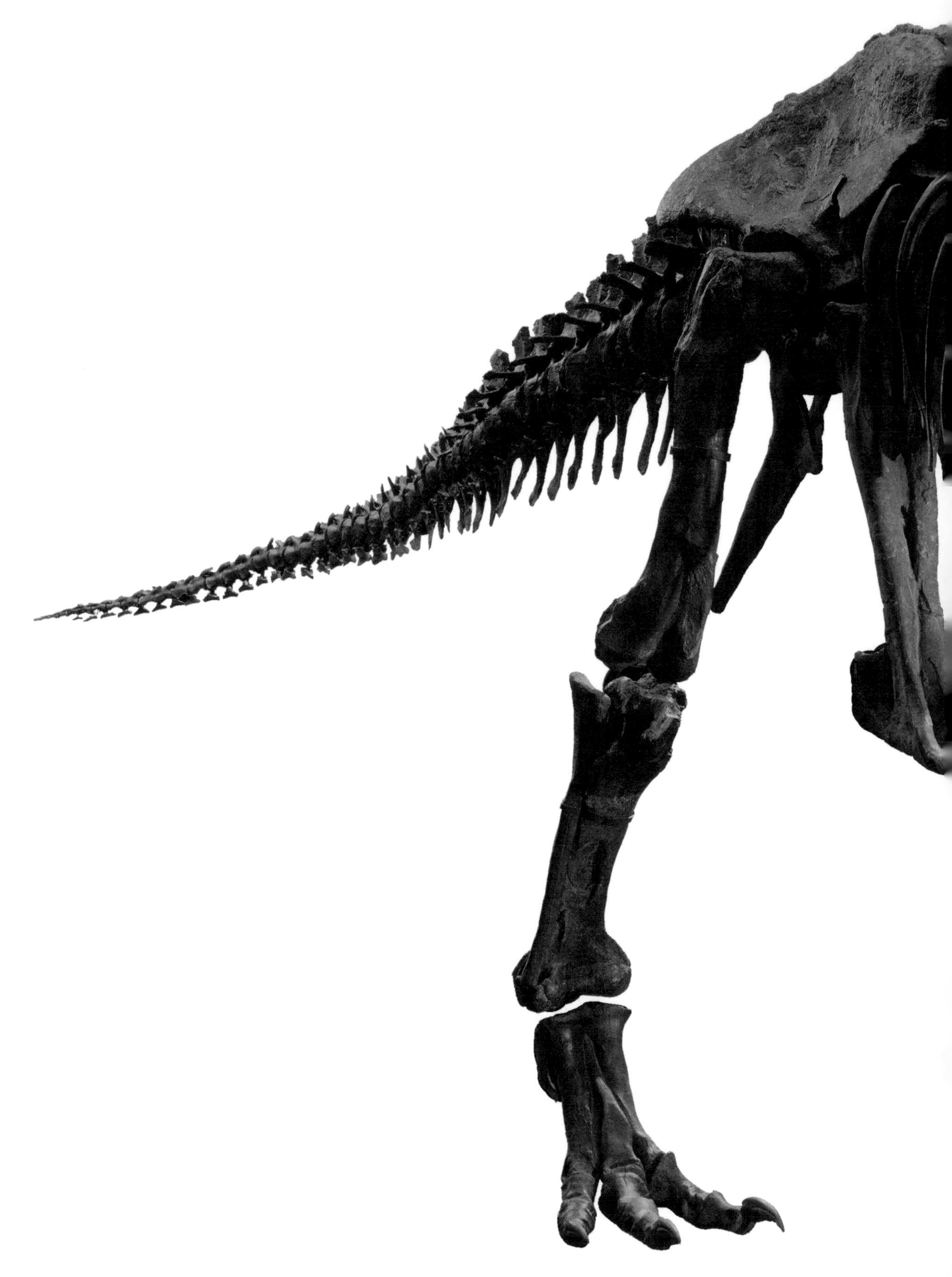

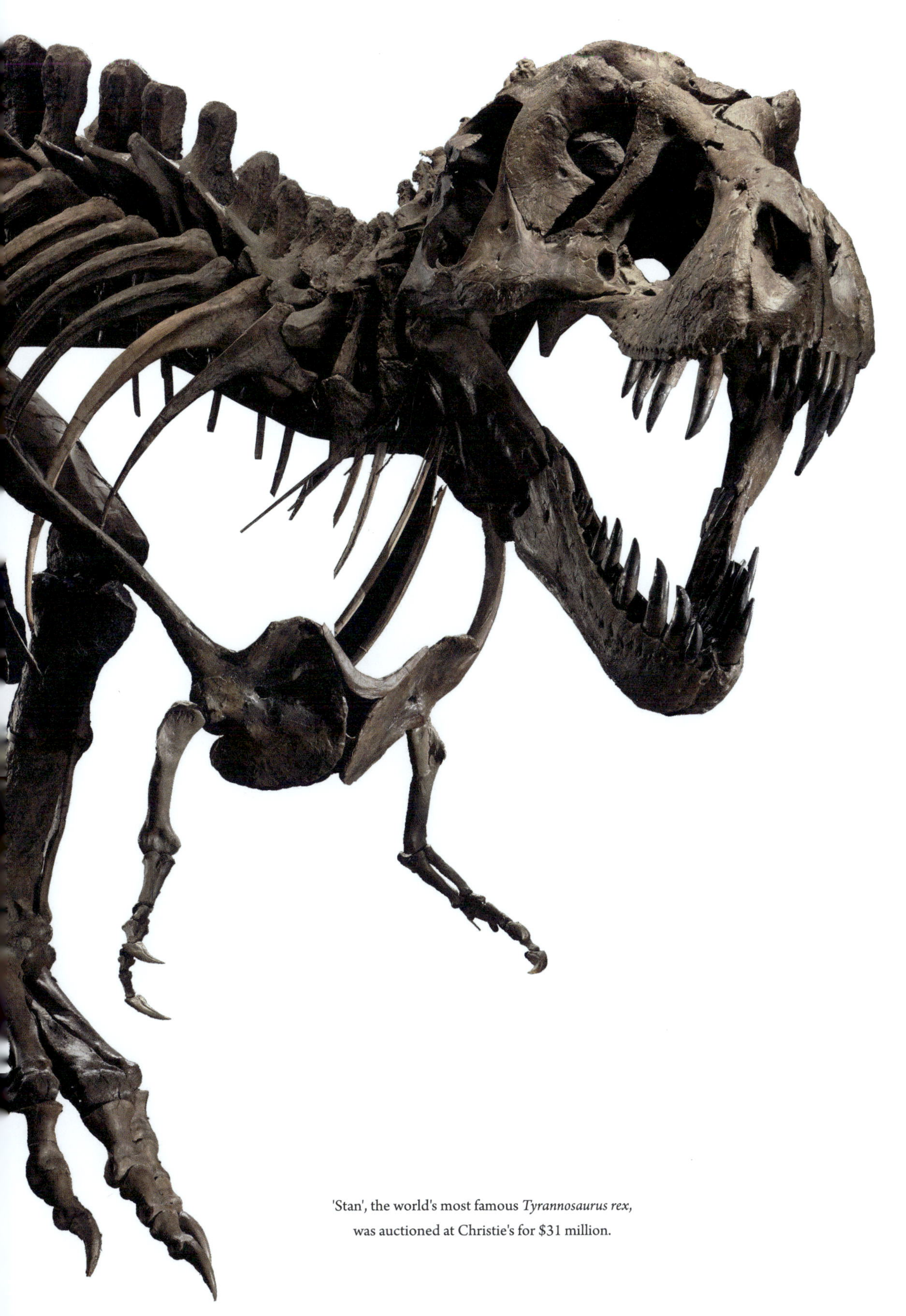

'Stan', the world's most famous *Tyrannosaurus rex*,
was auctioned at Christie's for $31 million.

CHECKLIST FOR DINOSAUR BUYERS

Seven things to look for when you buy dinosaur fossils:

1. Check the quality, origin and rarity.

2. Note how complete the skeleton is. Skeletons or skulls of common species can be up to 75 or 80 per cent complete. In the case of extremely rare species, 40 per cent is acceptable.

3. Buy a skeleton that accurately represents the creature. No one wants half a dinosaur. A missing tail vertebra isn't a problem. But what do you do with a dinosaur that has no hind legs?

4. Is the skeleton a 'Frankenstein amalgam'? Most dinosaur skeletons are assembled from the bones of different specimens, sometimes from completely different sites. And that's okay, but if the skeleton is an amalgamation, this ought to be stated transparently in the description. This is not a problem with mammoths, as the legs often wash ashore separately.

5. Check to see which parts of the skeleton are casts. It's common to replace missing bones with replica plaster fragments. If this is the case, it must be clearly stated in the description. A certificate that describes which parts belong to the original skeleton and which are reconstructions must accompany every sale.

6. Ask for photographs of the site, for documents approving the excavation and any papers relating to the fossil's exportation. The photographs of the site are important because they provide visual evidence of what the fossil looked like in the rocks and how the bones were dug out. This information shows what happened on site, and how the dinosaur was prepared and mounted afterwards.

7. A legal document providing proof of provenance is essential, given that some countries prohibit the exportation of fossils. However, under certain conditions, you are permitted to export bones from the United States, for example. But you need to take precautions when exporting fossils from countries such as Brazil, Mongolia or China, as American businessman Eric Prokopi discovered in 2012, when he wanted to sell a *Tarbosaurus* skeleton smuggled from Mongolia in New York. Actor Nicolas Cage also returned a skull of a *Tarbosaurus bataar* in 2015, because it came from Mongolia. In 2020, discussion arose about a spectacular Brazilian fossil of *Ubirajara jubatus* that ended up in a German institute under questionable circumstances, resulting in scientific research and publications about the fossil being frozen.

THE 'SUE' CONTROVERSY

Stan was not the first intact *T. rex* to be auctioned. In 1997, 'Sue' went under the hammer and, even then, sold for a staggering €8.7 million. Sue is one of the best-preserved tyrannosaurs ever discovered. But the remains are steeped in scandal, which was also the subject of the 2014 documentary *Dinosaur 13*. In 1992, the skeleton was seized from the Black Hills Geological Institute, where it was being studied after the fossil hunter Sue Hendrickson found it in 1990. A years-long legal battle ensued with Maurice Williams, the owner of the land where Sue was allegedly excavated. Williams won the case and regained possession of the skeleton, after which he decided to auction it at Sotheby's in 1997. The auction was as hair-raising as the chase scene in *Jurassic Park*. Bids started at a cautious $500,000. But because museums feared the priceless remains would end up in the hands of a private collector, frenzied fundraising had taken place behind the scenes beforehand. The Smithsonian was prepared to pay $2.5m. But in the end, the Field Museum of Natural History in Chicago bagged the spoils for over $8 million, with the financial support of private philanthropists, boosted by funding from McDonald's and Walt Disney. It was immediately decided to make several replica casts of the unique skeleton, so it could travel the globe.

It's worth noting that it was these very companies that helped purchase the fossil. Let's not forget, dinosaurs hold a unique place in our hearts as pop culture icons and are fabulously mediagenic. And, in that sense, Sue's auction was perfectly timed. In the 1990s, the *Jurassic Park* movie and its sequels unleashed a veritable dinomania. But dinosaur madness was never only for children and teens. Even famous film stars such as Leonardo DiCaprio, Nicolas Cage and Russell Crowe collect dinosaur bones, as do several other Hollywood figures. 'What happened after the Jurassic Park movies was that every wealthy person in the world apparently decided they had to have a dinosaur in their living room,' said palaeontologist Hans Sues, connected to the National Museum of Natural History at the Smithsonian Institution.

Auction-world insiders point out that, in recent years, a new generation of collectors from Japan, Singapore, Hong Kong, Korea and the United States has emerged, eager to buy these important dinosaur fossils, sometimes as an investment. Prehistoric remains find their way into private collections, private museums or public places, as tourist eye-catchers. 'Complete dinosaur skeletons have become a status symbol. As a timeless artifact, they will never date. Over time, the remains may be interpreted differently, but a skeleton such as this is timeless,' says British palaeontologist Mark Witton. He compares a complete skeleton to a classic car: 'You don't want a vintage vehicle that's been patched up, do you? You just want the original.'

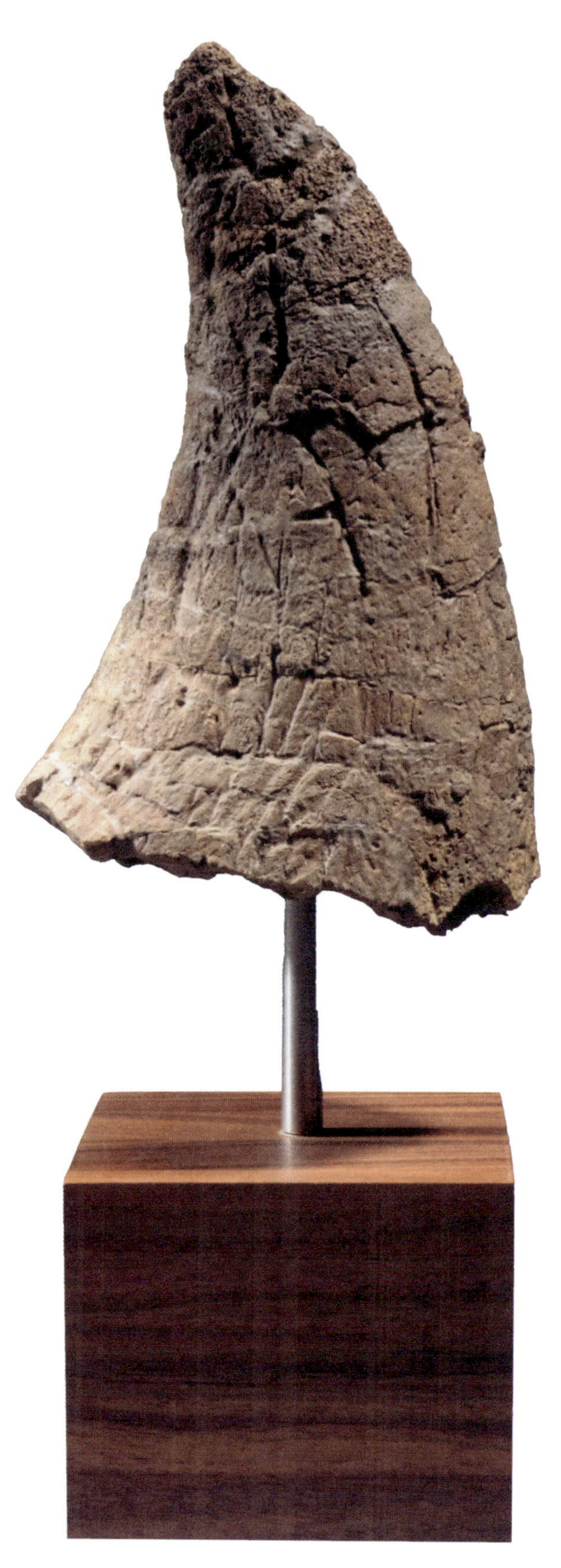

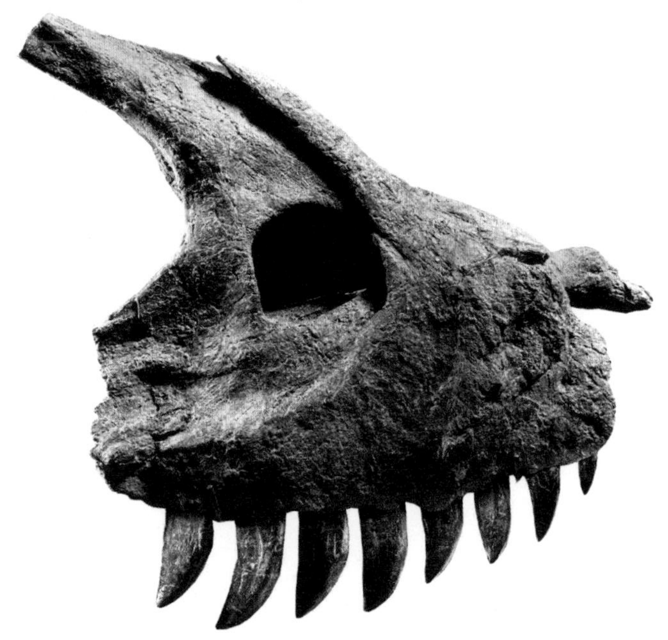

DINOSAUR FOSSILS IN THE *WUNDERKAMMER*

'I understand the global appeal: a dinosaur is more universal than a Picasso. And easier to appreciate; you don't need any formal training. Put bluntly, it doesn't require any "taste" on your part, just a fascination with dinosaurs and a dash of nostalgia for your childhood,' says Luca Cableri. The Italian is one of the few art dealers who regularly sell dinosaur fossils, both in their gallery and at art fairs. In his gallery in Arezzo, Cableri has already offered a complete *Allosaurus* and a *Diplodocus* for sale and, at the Brussels art fair BRAFA, sold a *Triceratops* skull. Cableri doesn't simply deal in fossils or natural history. Just like Christie's, he adopts a multidisciplinary approach, combining dinosaur remains with other collectibles reminiscent of the age-old *Wunderkammer* tradition, or 'cabinet of curiosities'. Think stuffed animals, ancient navigational instruments, exotica, and meteorites nestling among Hollywood props and even memorabilia from the history of space travel – the fossils of the future.

∧

The upper jaw of an *Allosaurus*.

<

The horn of a *Triceratops*, auctioned at Binoche et Giquello for €5410.

>

The Theatrum Mundi gallery in Arezzo, a *Wunderkammer*
that sells curiosities from prehistory until now.

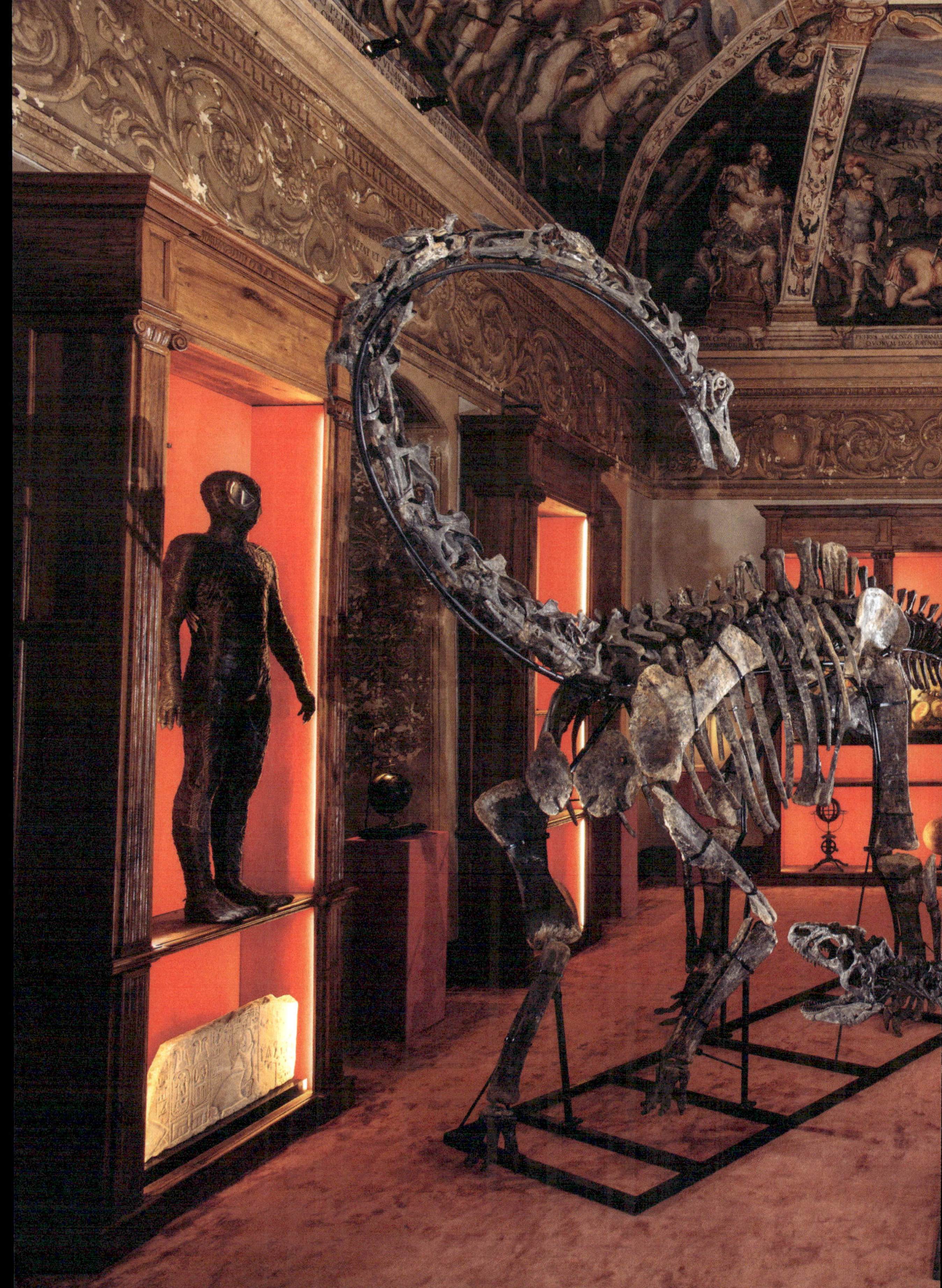

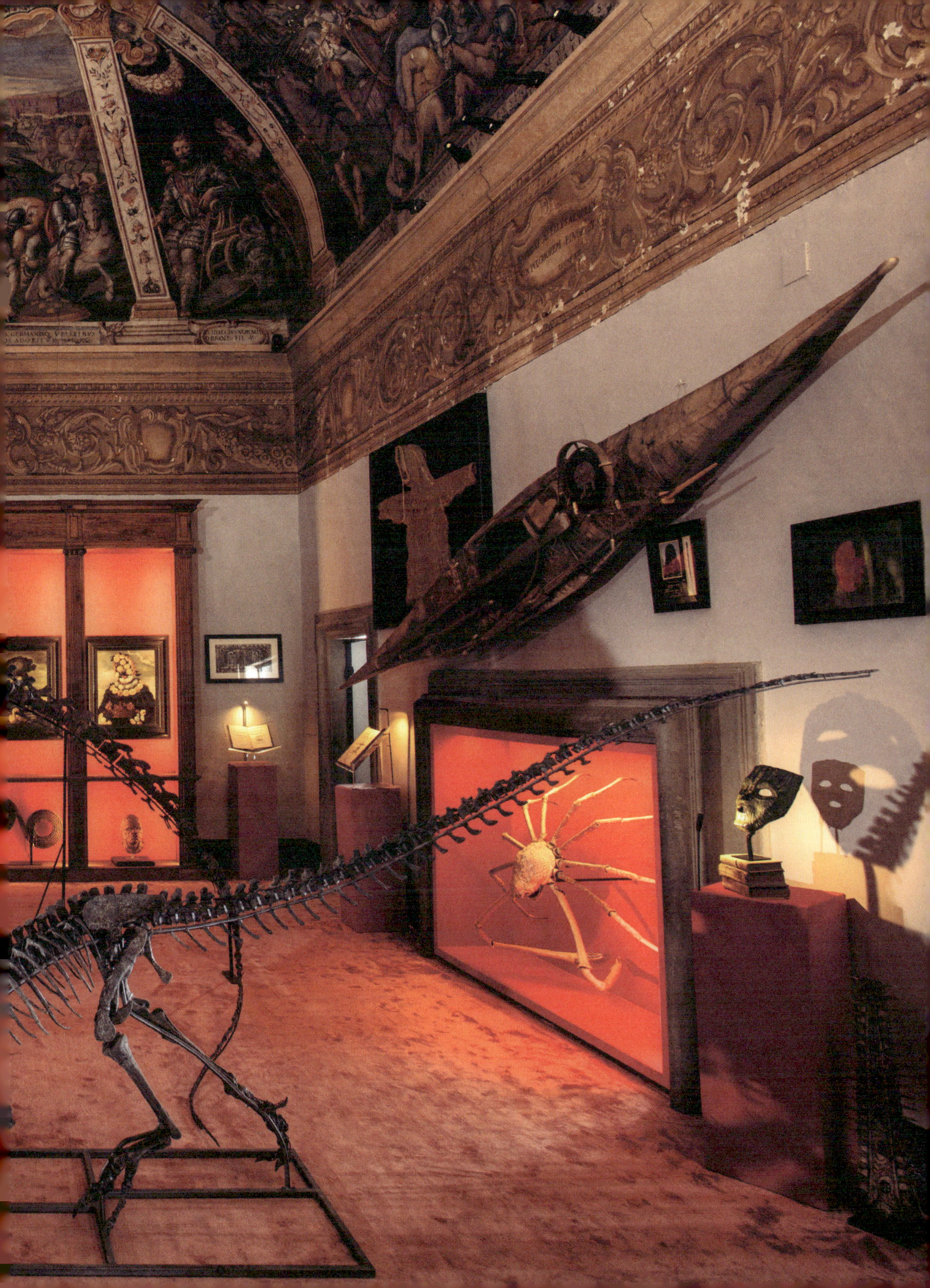

A NEW DINOSAUR UNDER THE EIFFEL TOWER

In recent years, dinosaur fossils reached a new group of private buyers, both collectors and speculators. For museums, this poses a potential problem, as they risk missing out on valuable specimens unless they have the financial support of companies like McDonald's or Walt Disney. Given that most museums simply can't afford to buy fossils at auction or from specialised dealers, they pursue other options. Consequently, many museums partner with private collectors to arrange loans so that a larger public can enjoy their purchase. For example, 'Arkhane', a skeleton bought by a private collector for €2 million at the French auction house Aguttes in June 2018, was on display in the Brussels Museum of Natural History at least until early 2022. Commotion was rife in the dinosaur world when, with great showmanship, the fossil went to auction in the Eiffel Tower. The Society of Vertebrate Palaeontology, of which some 2,200 palaeontologists worldwide are members, wanted the sale cancelled because the find was of great scientific value. The dinosaur turned out to be an unidentified species of *Allosaurus*. The palaeontologists were concerned that, if such a find were to be bought by a private collector, the fossil would be lost to science. After all, the so-called 'holotype' of a new dinosaur species should remain available to scientists for ever. The sale painfully exposed a common trend: dinosaur collectors work directly with middlemen rather than with museums. The Society of Vertebrate Palaeontology's last-ditch attempt to halt the sale failed. A private collector snapped up the remains. However, the palaeontologists' concerns were silenced when that private collector lent the fossil to the Institute of Natural Sciences in Brussels. This was good for science and for the public. 'The agreement was that Arkhane would be suitably displayed in our building, in exchange for expert research,' says Pascal Godefroit of the Royal Belgian Institute of Natural Sciences in Brussels. 'We agreed because we had certainty regarding the quality and traceability of the skeleton.' Arkhane not only attracted the attention of dinosaur lovers; the young contemporary artist Felix De Clercq immortalised the new dinosaur in a painting (see Chapter 4).

>

Kléber Rossillon's *Allosaurus* display case in the public gardens of Marqueyssac.

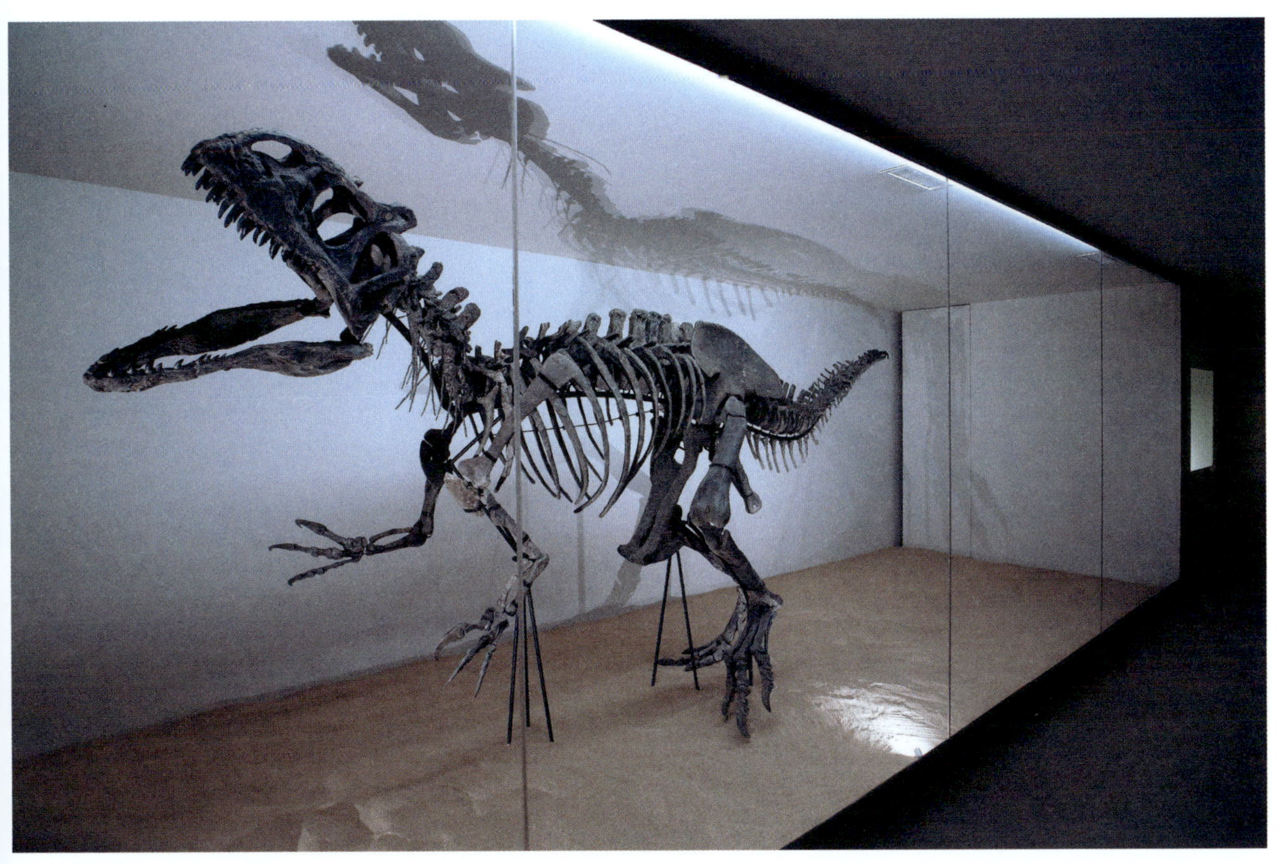

THE HEATHROW *DIPLODOCUS*

Dinosaur skeletons are a major attraction for visitors – a fact which French entrepreneur Eric Mickeler, co-founder of Thétis, a company that provides dinosaur skeletons to institutions and museums worldwide, is well aware of. In 2009, he collaborated on a huge exhibition at the Grand Palais in Paris, where the stars of the show included a *Stegosaurus*, *Allosaurus*, *Plesiosaurus* and about 50 other fossils. He is, also the driving force behind the newly opened Dinosaur Museum in Prague, and arranged for the giant *Diplodocus* 'Skinny' to be installed in London Heathrow's departure terminal in 2019. Mickeler was also involved in the sale of an *Allosaurus* in the Eiffel Tower in 2017. Kléber Rossillon, a French entrepreneur who invests in tourist attractions, bought the specimen for €1.12 million. The *Allosaurus* has been on display in a glass box in his hanging gardens of Marqueyssac (in France's Dordogne region), since 2017. Rossillon has installed the skeleton as a prehistoric attraction, a majestic creature gazing out over the romantic topiary. Other companies have also recently invested in dinosaur skeletons as eye-catchers in their corporate headquarters. Swiss company Novartis has an *Allosaurus* skull. Eximium bought the skull of a *Triceratops*. And the Belgian serial entrepreneur Marc Coucke bought Skinny, as well as a *Stegosaurus*, for his theme park Adventure Valley in Durbuy.

ITINERANT DINOSAURS

The German entrepreneurial family Pohl have also invested in a world-class collection of fossils and minerals. They conduct palaeontological research themselves and also excavate dinosaurs. Their most important finds are on display at their Wyoming Dinosaur Center and at their Sino-German Palaeontology Museum in China. The Pohl family also lends important items from the collection for scientific research, for example to museums, universities and research institutes. In collaboration with the American Granada Gallery and the Belgian jeweller Jochen Leën, collection pieces – dinosaur fossils and replicas – are sometimes lent for major events. During the TEFAF Maastricht 2019 art fair, Leën held a dinosaur exhibition in the luxury hotel La Butte aux Bois, with a *Camptosaurus* skeleton as the ultimate eye-catcher. During Art Basel in Miami, he helped organise a dinner where guests could eat among dinosaur skeletons, a reference to the legendary 'dinner at the *Iguanodon*' that Benjamin Waterhouse Hawkins hosted on New Year's Eve in 1853 (see Chapter 2).

KEEP AN EYE OUT

Although dinosaur fossils frequently come up for sale at auction houses, online auctions, galleries, second-hand sites, dealers and art fairs such as TEFAF or BRAFA, you need to be on guard. Many skeletons are composed of various animals, and some have been over-restored. 'There is a fairly small legal market in which buyers collaborate with museums. And then there are fossils sold on the black market, which originate from plundered sites or locations where excavating is illegal,' says natural history specialist Iacopo Briano. 'You need to keep an eye out, because most fossils are being sold by traders working illegally,' he warned in the art magazine *Collect*.

>

The Granada Gallery is a specialist dealer of (dinosaur) fossils,
skeletons and rare minerals.

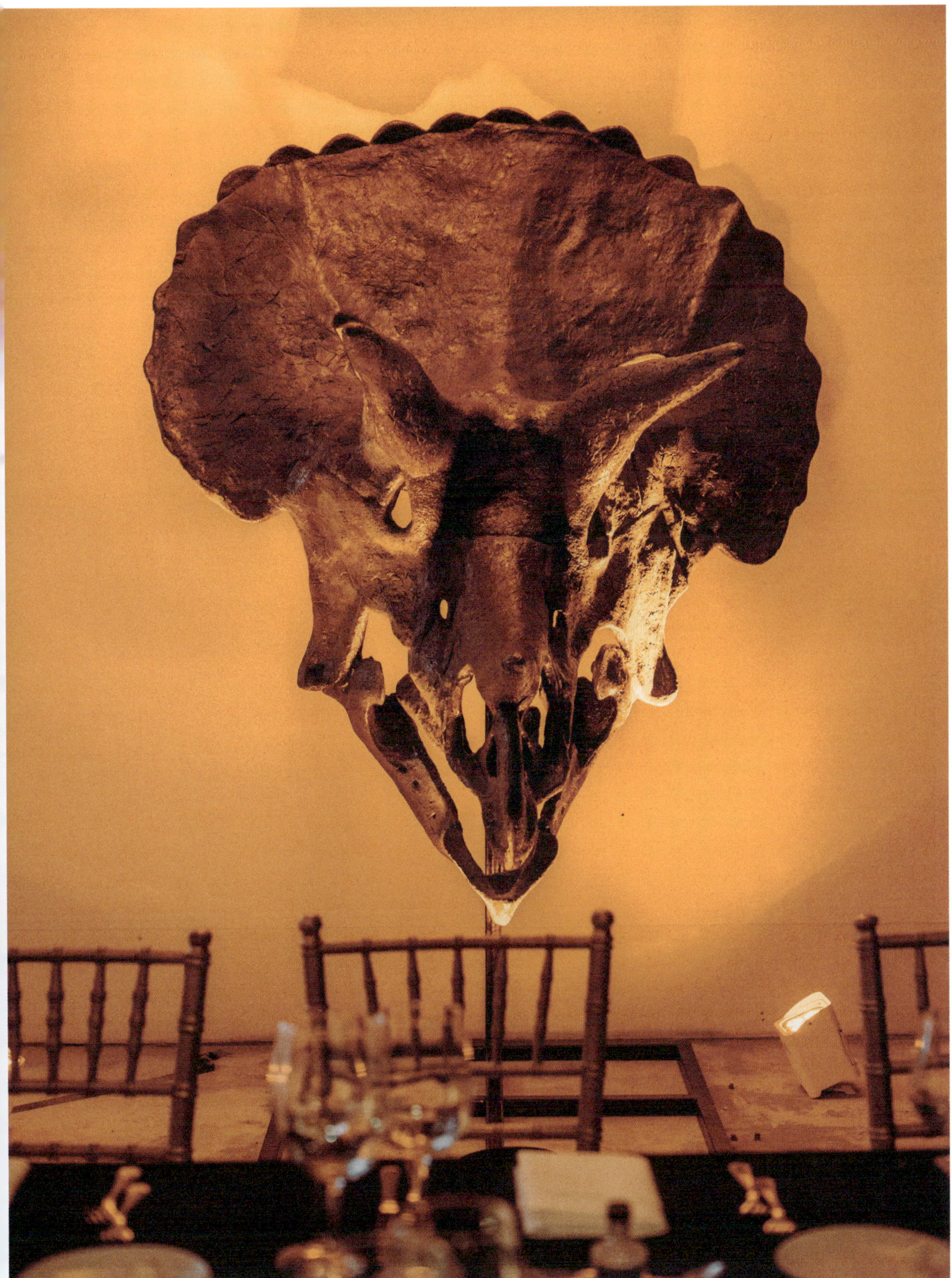

5

UNWRITTEN RULES
FOR DINOSAUR FOSSIL COLLECTORS

1. Carnivores are more sought after than herbivores because many people (mistakenly) associate dinosaurs with bloodthirsty reptiles.

2. Baby dinosaurs are more sought after than adults because of their rarity value.

3. Skulls sell more easily than intact dinosaur skeletons which require considerable space.

4. Skeletons in perfect condition are extremely rare, and thus the most expensive. But they do come up for sale occasionally at specialised galleries and at auction houses.

5. Watch out for so-called 'fauxiles'. Like every branch of the arts that can be monetised, there are many dinosaur fossil fakes on the market. So, be alert and consult an expert if necessary.

>

The *Giraffatitan brancai* versus a human skeleton in 1937
at the Museum of Natural History in Berlin (see p. 72).

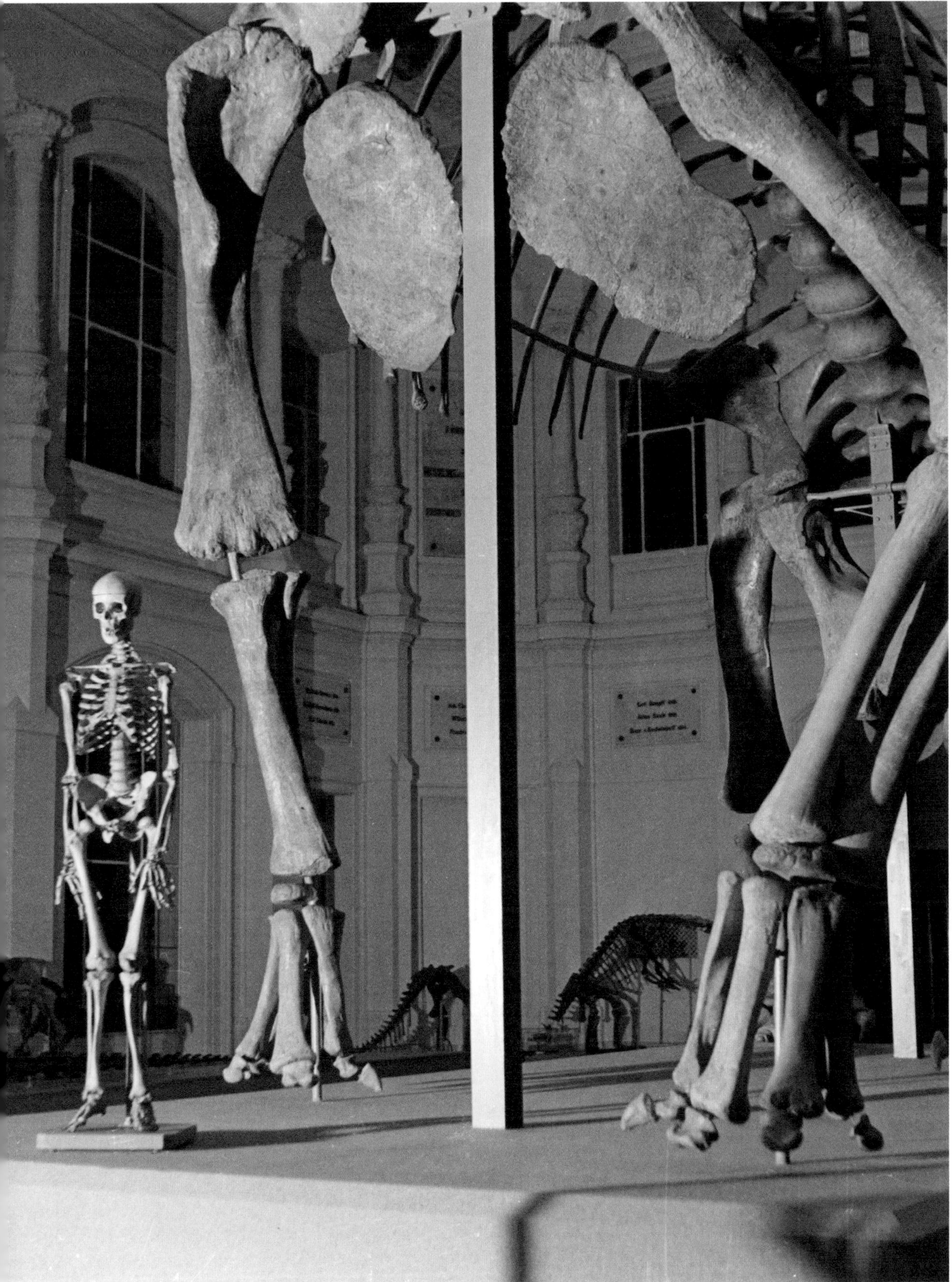

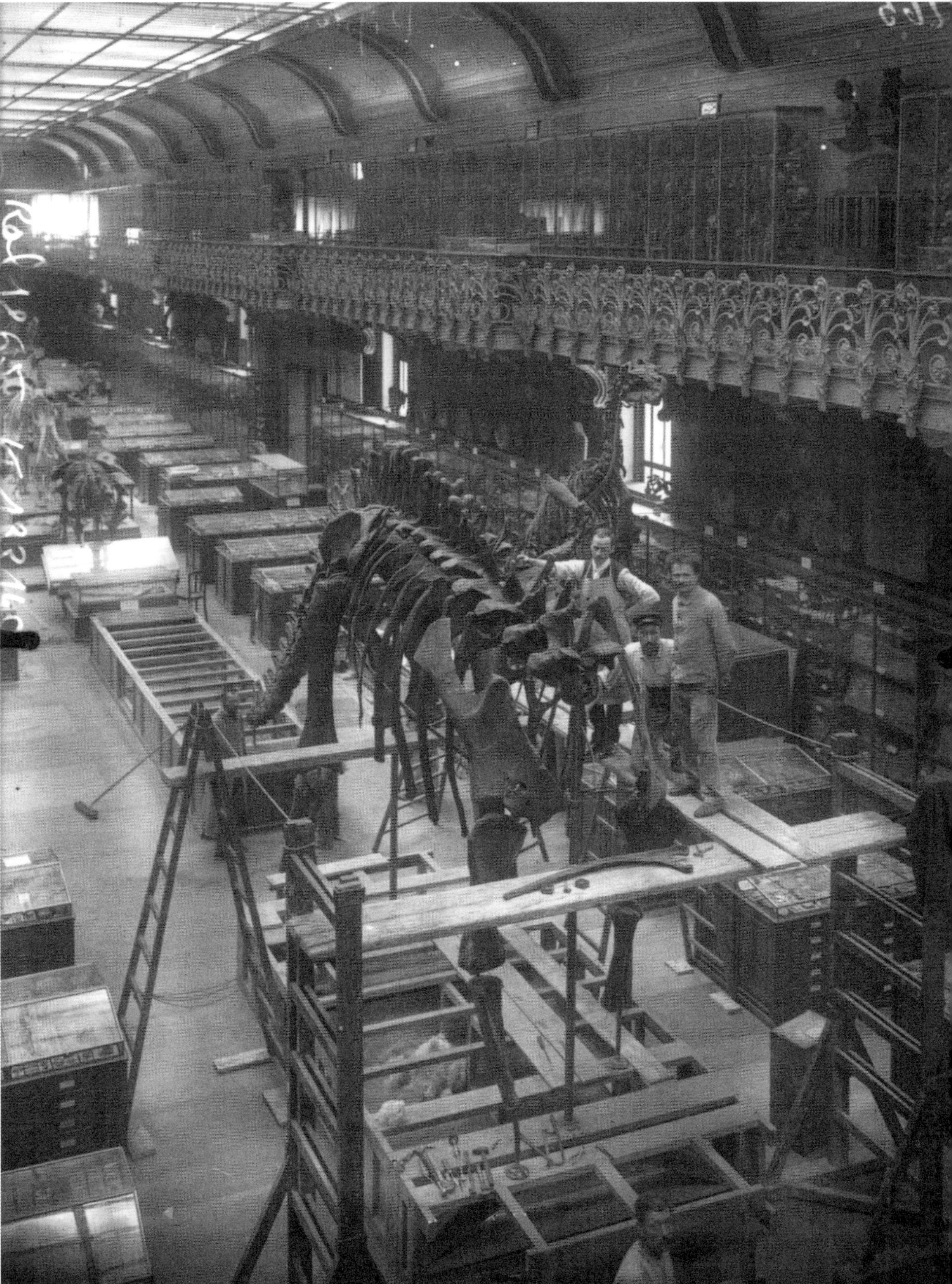

DIGGING FOR DINOSAURS

DISCOVERING AND NAMING DINOSAURS

'Dinosaurs came into public consciousness in the same period and as a product of the same forces that produced the tank, the locomotive, the steamboat, and the skyscraper. The cutting open of the Western landscape by the railroads was spilling dinosaur bones out of their primeval graves. The railroads were bringing back carloads of bones to New York, Philadelphia, and New Haven. And railroad money was helping to finance the bone rush and the new museums.'

The Last Dinosaur Book, W.J. T. Mitchell (1998)

<

The *Diplodocus* replica under construction
in the Musée Nationale d'Histoire Naturelle, Paris.

LIKE ATTRACTS LIKE

Many people instantly think of dinosaurs as monstrous creatures. The largest were 40m long and weighed up to 80 tonnes, like the *Patagotitan* or the *Argentinosaurus*. But the smallest were barely the size of a chicken. They ruled our planet for 150 million years until they suddenly became extinct about 66 million years ago. There are countless types of dinosaur – so many that you may wonder what exactly a species is. If two individuals can produce fertile offspring, then they belong to the same species. At least, that's the biological definition, one that is difficult to test on fossils of extinct animals, of course. Which is why palaeontologists adopt a morphological concept of species. They describe a skeleton's anatomical features and try to find out whether these sufficiently different from those of other skeletons. This allows them to determine whether a found skeleton belongs to a new or a known species. Of course, every palaeontologist dreams of discovering their very own dinosaur species. And in the best-case scenario, of having one named in their honour.

WHAT'S IN A NAME?

You can't just come up with a name for a dinosaur species. Like biologists, palaeontologists use taxonomy to name their species. There is an International Code and a Commission for Zoological Naming for this purpose. Names are usually derived from Latin or Greek, although there is room for some poetic licence. The name often refers to the location. This is the case with the *Ampelosaurus atacis*, named after the southern French vineyards, where it was discovered: *ampelos* is Greek for 'vine'; *Atax* is Latin for the River Aude near which it was excavated. The Commission for Zoological Naming checks whether the names do not already exist, but also that they make sense and do not conflict with other names. For example, the very first dinosaur fossil described in 1763 was called *Scrotum humanum*. It did indeed resemble a fossilised human scrotum. The fossil itself has been lost, although it most likely belonged to a *Megalosaurus*. In the end, the name didn't stick. The first official dinosaur was *Megalosaurus bucklandii*, named after its discoverer, William Buckland (1784–1856).

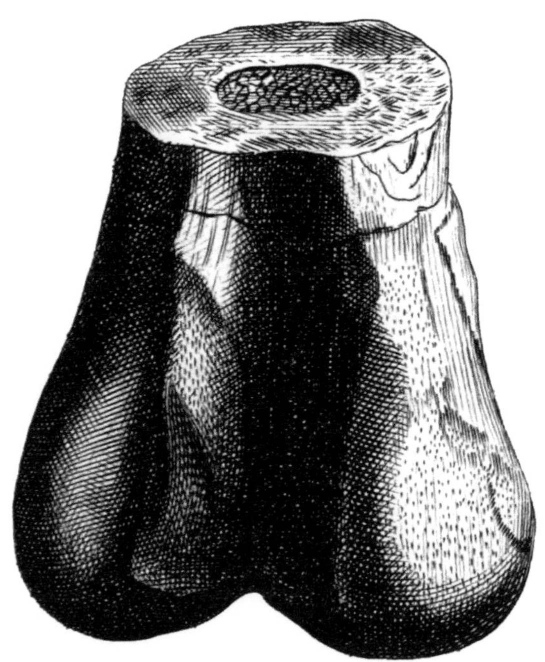

The very first dinosaur bone described in 1677 was nicknamed *Scrotum humanum* in 1763.
It was later found to have come from a *Megalosaurus.*

MOHAMMADISAURUS

A dinosaur name can also change over time – sometimes because science comes up with new insights, sometimes because the zeitgeist changes. When German palaeontologists carried out an expedition in the German colony of German East Africa (today's Tanzania) between 1909 and 1913, they were assisted by the local population, not only during excavations but also when transporting the fossils to the harbour. Out of respect for the hard work of their African helpers, the German palaeontologists named the species of dinosaurs they found after them. They chose names like *Mohammadisaurus*, *Abdallahsaurus* and *Selimanosaurus*. One dinosaur type was even named *Wangonisaurus*, after the Ngoni, the people to which many of the workers belonged. These informal names were a sign of respect and recognition for the African assistants. 'The people work more quickly than I had expected, but also at the same time quite carefully. They also have a good eye for what is bone and what is rock. Some of them understand how to expose with the greatest care and exactitude bones as difficult as ribs,' we read in the German fieldnotes of the time. But despite the assistants' valued (and paid) services, the African names quickly disappeared when the fossils arrived in the Museum für Naturkunde in Berlin. There, the informal names were replaced by 'common' taxonomic names, which often featured the names of the German discoverers. The museum is aware of the colonial problems surrounding this naming issue. When palaeontologist Kristian Remes discovered a new dinosaur type in 2007 on the basis of Tanzanian fossils, it was named *Australodocus bohetii* in honour of the Tanzanian chief preparator, Boheti bin Amrani, who played an important role in the Germans expeditions. It was the first time a Tanzanian had appeared in the official naming ceremony. The same thing happened in 2019. The *Wamweracaudia keranjei* is named after the Wamwera, a Tanzanian language group to which local fossil expert Mohammadi Keranje belongs.

THE HOUSE SPARROW AND
THE DINOSAUR FAMILY TREE

Dinosaurs cannot be defined by one particular characteristic. That would be almost impossible, given their great diversity. Scientists therefore use the term 'clade' to define the Dinosauria group. Animals belong to the same clade if they share a range of traits and originate from a common ancestor. This is not unimportant from an evolutionary perspective. The clade of dinosaurs is, after all, described as 'all the descendants of the last common ancestor of *Triceratops horridus* and *Passer domesticus* (the house sparrow)'. That sounds complicated, but if you draw it on a family tree (cladogram), this family tree encompasses all the animals we describe as dinosaurs today. All descendants of this common ancestor belong to the Dinosauria group.

4
DINOSAURS WITH BIZARRE NAMES

1. ***Dracorex hogwartsia*** is a recently named type of dinosaur, found in 2003 and named after Hogwarts School of Witchcraft and Wizardry, school for magical studies in J. K. Rowling's Harry Potter stories. Its bony head, full of lumps and bumps, is indeed reminiscent of a dragon.

2. ***Irritator challengeri***: After illegal excavations in Brazil, a promising fossil ended up in the Staatliches Museum für Naturkunde in Stuttgart. Researchers thought they had discovered a new species of pterosaur, but the skull turned out to have been heavily damaged and altered. The species was named *Irritator* because the research team was so irritated about the falsification. The *Irritator challengeri*, for the time being the only known species *of Irritator*, is a carnivore related to the *Spinosaurus*. The species name refers to Professor George Edward Challenger, the main character in Arthur Conan Doyle's book *The Lost World* (1912) – a palaeontologist who is on a quest to discover a remote location on a high plateau in Central America where prehistoric animals were believed to have survived.

3. ***Gasosaurus***: A carnivorous theropod, discovered at a construction site in China where the authorities wanted to build a gas terminal. When attempting to blow up the substrate, workers unearthed a skeleton that was badly damaged during the explosion. For years, scientists had known that the site's subsoil was rich in fossils, but the Chinese government hadn't wanted to jeopardise the construction of the gas plant in the interests of scientific research.

4. ***Bambiraptor***: In 1993, 14-year-old fossil hunter Wes Linster discovered the skeleton of a carnivorous theropod in Montana that was so small that researchers eventually named it after the Disney character Bambi. The name literally means 'Bambi thief'.

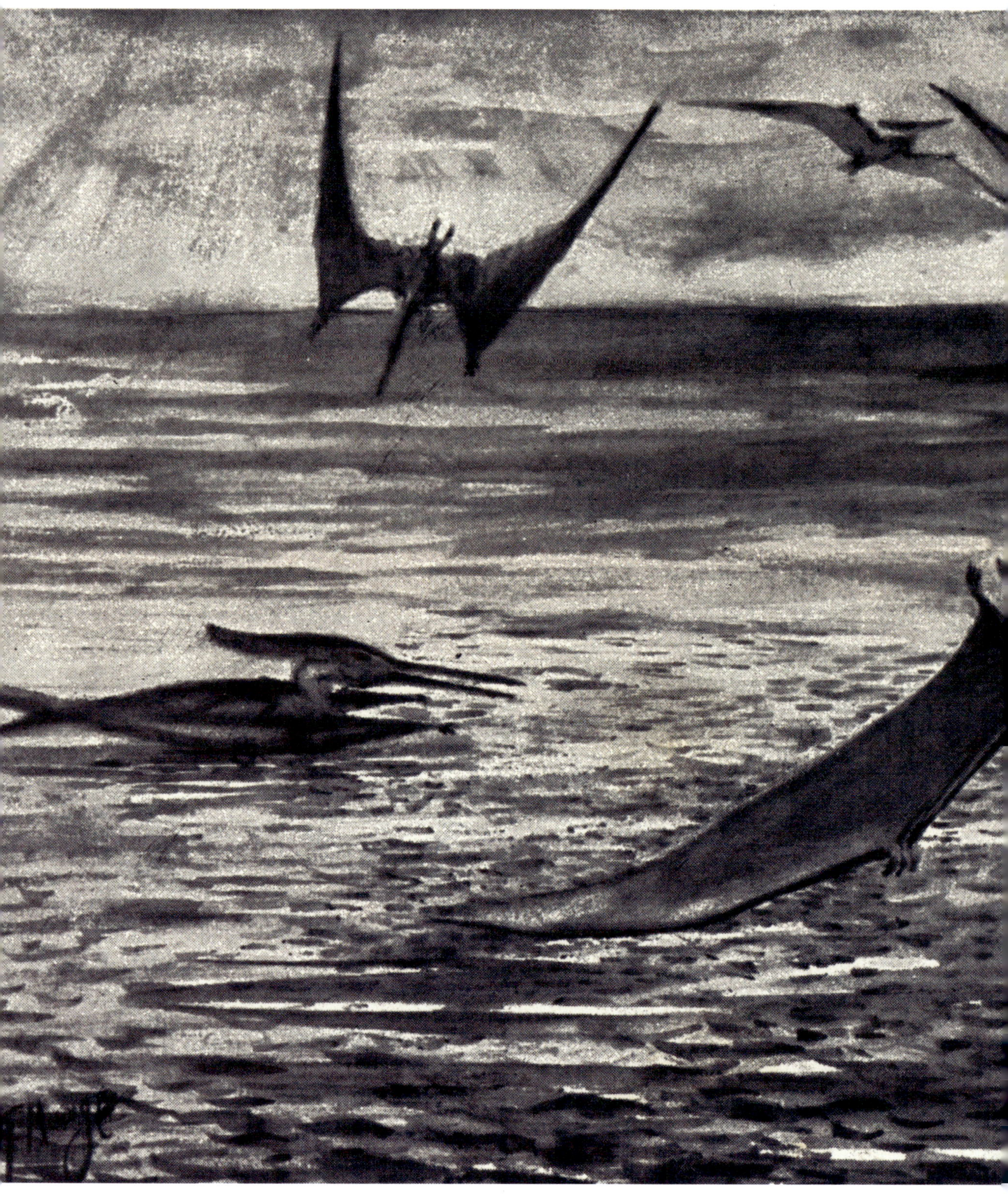

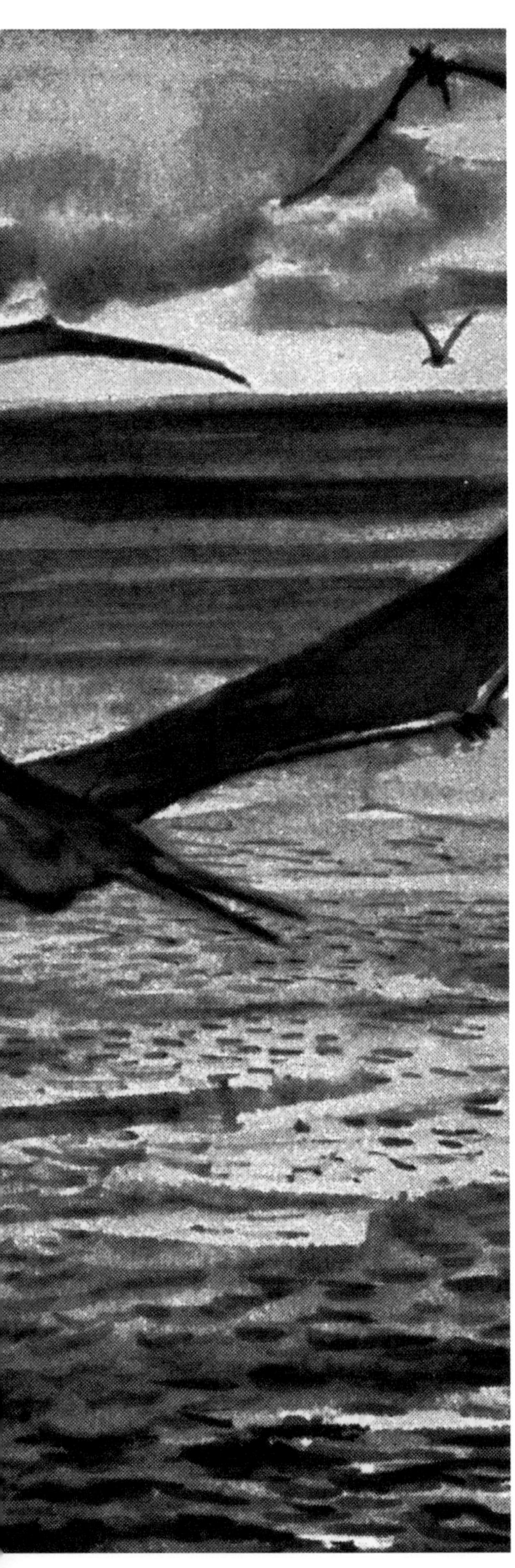

ARE BIRDS THE DINOSAURS OF TODAY?

The main feature that distinguishes dinosaurs from other reptiles is that their legs are directly under their bodies. Experts call this position 'graviportal'. In other reptiles, the legs stand sideways, the so-called 'parasagittal' position. Think of crocodiles, lizards and turtles. This means that reptiles that swam in the sea during the time of the dinosaurs (ichthyosaurs, plesiosaurs, mosasaurs) are not dinosaurs. And that reptiles that flew in the air in those days (pterosaurs, for example) cannot officially be called dinosaurs either. Birds are the only flying dinosaurs. They evolved from a group of smaller theropods (carnivorous dinosaurs), the Maniraptora, which had an integral plumage. The feathers not only kept the small animals warm in colder regions, but were also useful as wings during acrobatic jumps. Some of these prehistoric birds also moved very differently from the modern birds we know. A *Microraptor* had long feathers on its hind legs, and with its four wings it could soar through the air. Unfortunately, only modern birds survived the mass extinction at the end of the Cretaceous Period, about 66 million years ago. This means that dinosaurs never really became extinct. The avian dinosaurs, or birds, as we know them, are their direct descendants.

———

Hugo Wolff-Maage's illustration of pterosaurs – strictly speaking, not dinosaurs – for Wilhelm Bölsche's 1910 book *Das Leben der Urwelt* (Life of Primeval World).

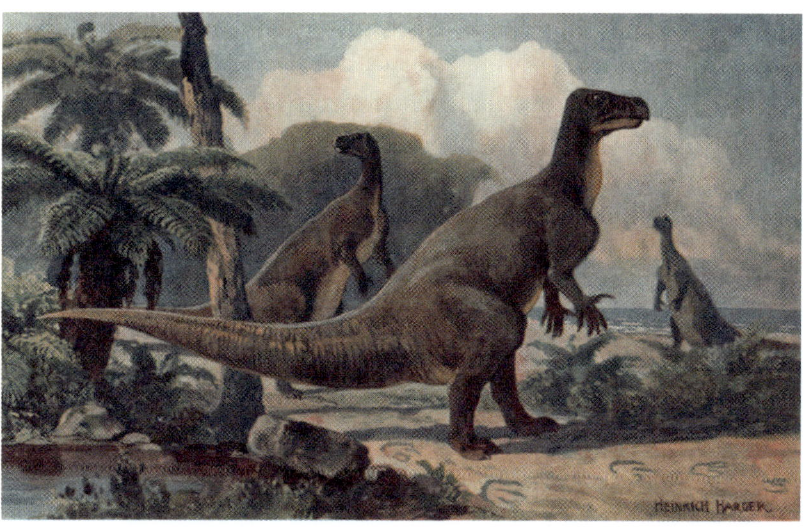

Iguanodons, drawn in 1916 by book illustrator Heinrich Harder.

FROM 230 TO 66 MILLION YEARS AGO

The word 'dinosaur' usually brings to mind the typical large extinct reptiles, whose unpronounceable names many children remember best. To refer to this specific group, scientists use the English term non-avian dinosaurs. Most scientists recognise two large groups, which differ from each other mainly in the shape of their pelvises. The Ornithischia, to which animals such as *Stegosaurus, Iguanodon, Triceratops* and *Ankylosaurus* belong, have a pelvis in which the pubic bone is turned backwards and runs parallel to the ischium. The name Ornithischia literally means 'with a bird's ischium'. But ironically, birds evolved from the other group, the Saurischia. Saurischia have a pubic bone that points forward, at a distinct angle to the ischium. This group includes two large branches, on the one hand the Sauropomorpha (sauropods) or long-necked dinosaurs, with animals such as *Plateosaurus, Diplodocus* and *Patagotitan,* and on the other the Theropoda which eat almost exclusively meat, with animals such as *Velociraptor* and *Tyrannosaurus.* These groups all have a very complex evolution that took place from about 230 to 66 million years ago, and which in the case of the birds continues even today. The non-avian dinosaurs lived in an era that geologists call the Mesozoic. The Mesozoic is in turn divided into three major periods, the Triassic (252 to 201 million years ago), the Jurassic (201 to 145 million years ago), and the Cretaceous (145 to 66 million years ago).

WHICH WAS THE FIRST DINOSAUR
TO WALK THE EARTH?

'One thing is for sure: the dinosaurs originated during the Triassic period. Nearly everything else is uncertain,' writes palaeontologist Michael J. Benton in his book *The Dinosaurs Rediscovered*. The first three dinosaurs discovered and named were *Megalosaurus* (1824), *Iguanodon* (1825) and *Hylaeosaurus* (1833). The question which dinosaurs were the very first to walk the Earth is less easy to answer, though. For a long time, there was much uncertainty surrounding the origin of dinosaurs because there were so few fossils from the early period. After the first brief descriptions of *Eoraptor* in 1993 and *Saturnalia* in 1999, this field steadily gained momentum. Around 2010, findings caught up, with new species and more detailed descriptions of known species found in the Ischigualasto Formation and the Santa Maria Formation from the Middle Triassic of Argentina and Brazil, respectively. The species *Eodromaeus, Panphagia* and *Sanjuansaurus* joined *Eoraptor, Herrerasaurus, Saturnalia* and *Staurikosaurus* on the list of oldest dinosaurs, dating from between 233 and 231 million years ago. *Nyasasaurus* is possibly an even older dinosaur. It was found in Tanzania, in rocks that were 245 million years old. The remains were first described by British palaeontologist Alan J. Charig in 1956, but were not formally published until 2013, by Sterling Nesbitt and his team. Unfortunately, because of the find's fragmentary nature, it's impossible to be sure if the species falls just inside or outside the dinosaur family tree.

WHY IT'S *T. REX* AND NOT T-REX

Never say *Tyrannosaurus rex* to a *Tyrannasorus rex*. OK, it's only a two-letter difference. The first is the most famous carnivorous dinosaur ever, which lived between 70 and 66 million years ago. The second is a beetle found in amber from the Miocene (15 to 20 million years ago). Also, never write T. Rex, T-Rex or T-rex. The correct spelling is *T. rex*. Yes, in italics. Names of larger groups of dinosaurs, such as Maniraptora or Iguanodontidae, are always capitalised. The genus and species names are written in italics and lower case. If the name is used in a corrupted form, such as maniraptora or iguanodontidae, write it without a capital letter. If you have mentioned a genus and species name in full before in a text, you can also abbreviate the genus with the first letter and a full stop, followed by the species name, both again in italics. So *T. rex. U.nderstood*?

THE FIRST DINOSAURS WERE
NO BIGGER THAN A CAT

The first dinosaurs on Earth were not the gigantic bloodthirsty monsters that children usually imagine. In the Late Triassic, *Eoraptor*, *Saturnalia*, *Coelophysis* and *Herrerasaurus* roamed the planet. These animals were primarily carnivores, and were no bigger than a cat or dog. Towards the end of the Triassic, some sauropodomorphs began to assume larger sizes, such as *Plateosaurus* and the enormous *Ingentia prima*, which was first described only in 2019. In the Jurassic Period, sauropods (dinosaurs with long necks which walked on four legs) evolved into the largest animals ever to walk the Earth. The best-known examples are *Brontosaurus*, *Brachiosaurus* and *Supersaurus*. In the Late Jurassic Period (around 150 million years ago), other groups began to display greater anatomical diversity, such as *Camptosaurus* and *Stegosaurus*, both contemporaries of *Brontosaurus*. In the Cretaceous Period, the Ornithischia became increasingly widespread. New groups emerged with various spikes, horns, plates and beaks. Think of the horned ceratopsids, such as *Triceratops* and its relatives. Or the ankylosaurs with their bony armour and 20kg club-like tails, which could deal blows with a force of 2000 newtons. And, of course, the ornithopods, with *Iguanodon*, and later the hadrosaurs. The diversity of these different groups of dinosaurs is incredible, and we certainly haven't discovered all the species yet. The beauty of it is that nature has been able to experiment with evolution for millions of years. It is up to palaeontologists to reconstruct this vast evolutionary puzzle from fossils.

EARLY DINOSAUR DISCOVERIES

The first recorded discoveries and scientific studies of dinosaur bones were made in England. Robert Plot (1640–1696), a chemistry professor at Oxford, described the distal end of a femur in his *Natural History of Oxfordshire* in 1677. He realised the bone was too large to belong to any hitherto-known species. So he concluded that it was a human giant, perhaps a Titan or some other kind of mythical being. In 1763, the same bone was catalogued by the natural scientist Richard Brookes (1721–1763). Because it resembled male genitalia, he named the bone – as mentioned earlier – *Scrotum humanum*. It came close to being the name of the first dinosaur bone ever described. But the name was forgotten, and in 1824 *Megalosaurus bucklandii* officially became the first dinosaur species.

The Welshman Edward Lhuyd (1660–1709) succeeded Robert Plot as director of the Ashmolean Museum in Oxford. He illustrated several fossil teeth in his 1699 publication *Lithophylacii Britannici Ichnographia*. Two of them are definitely dinosaur teeth. One closely resembles the tooth of a *Megalosaurus*, the species described only 120 years later by William Buckland (1784–1856). Another looks similar to the tooth of a *Cetiosaurus*, a long-necked dinosaur. This would make his illustration the oldest artist's impression of a sauropod fossil. Lhuyd named the sauropod tooth *Rutellum implicatum*. He believed it to be the tooth of a fish, probably because, at that time, there were no known reptile fossils.

ABNORMAL CROCODILE

In the 18th century, a number of fossilised remains, which we now know belong to dinosaurs, were found and described. Most of these were found in England, but also in France and North America. In France, the priests Jacques-François Dicquemare and Charles Bachelay respectively described bones discovered in the Vaches Noires cliffs between Houlgate and Villers-sur-Mer and in the region around Honfleur, both in Normandy. Bachelay's finds were illustrated and described in 1808 by the famous natural scientist Georges Cuvier (1769–1832) in a treatise on crocodilians. They were later given the name *Streptospondylus altdorfensis*. Cuvier realised that these were not normal crocodilians, but it was not until 1887 that Lennier was able to demonstrate with certainty that these were the bones of a theropod dinosaur. In North America, Caspar Wistar and Timothy Matlack described a large femur bone in 1787, but like many other very old specimens it is now lost. The American finds from the 18th century have had little or no influence on the scientific recognition of dinosaurs.

THE FIRST *IGUANODON*?

From the 19th century onwards, scientists began to realise that some of these large bones had reptilian features. Great strides had been made in research, especially in England. In 1869, Harry Seeley (1839-1909), a professor at the University of Cambridge, described a series of bones that can still be seen in the Sedgwick Museum of Earth Sciences in Cambridge. There, they are attributed to the sauropod *Cetiosaurus*. Thomas Webster reported in 1824 that a number of large bones found on the Isle of Wight looked rather reptilian. These were probably the first fossils of *Iguanodon*, a species that is still regularly found there today. But, as is often the case, these old specimens have been lost. Whether it was the first-ever excavated *Iguanodon* we will never know.

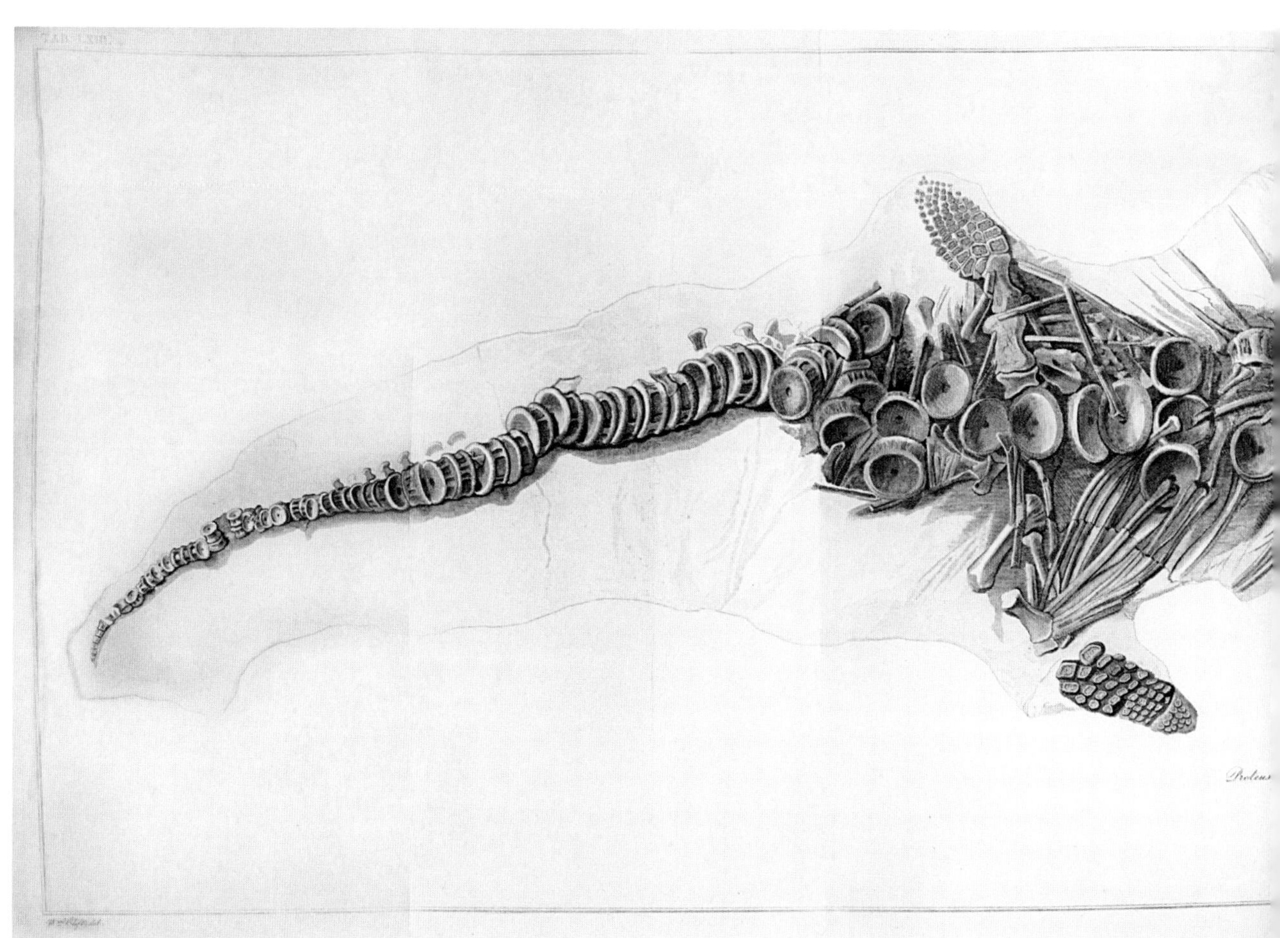
Proteus

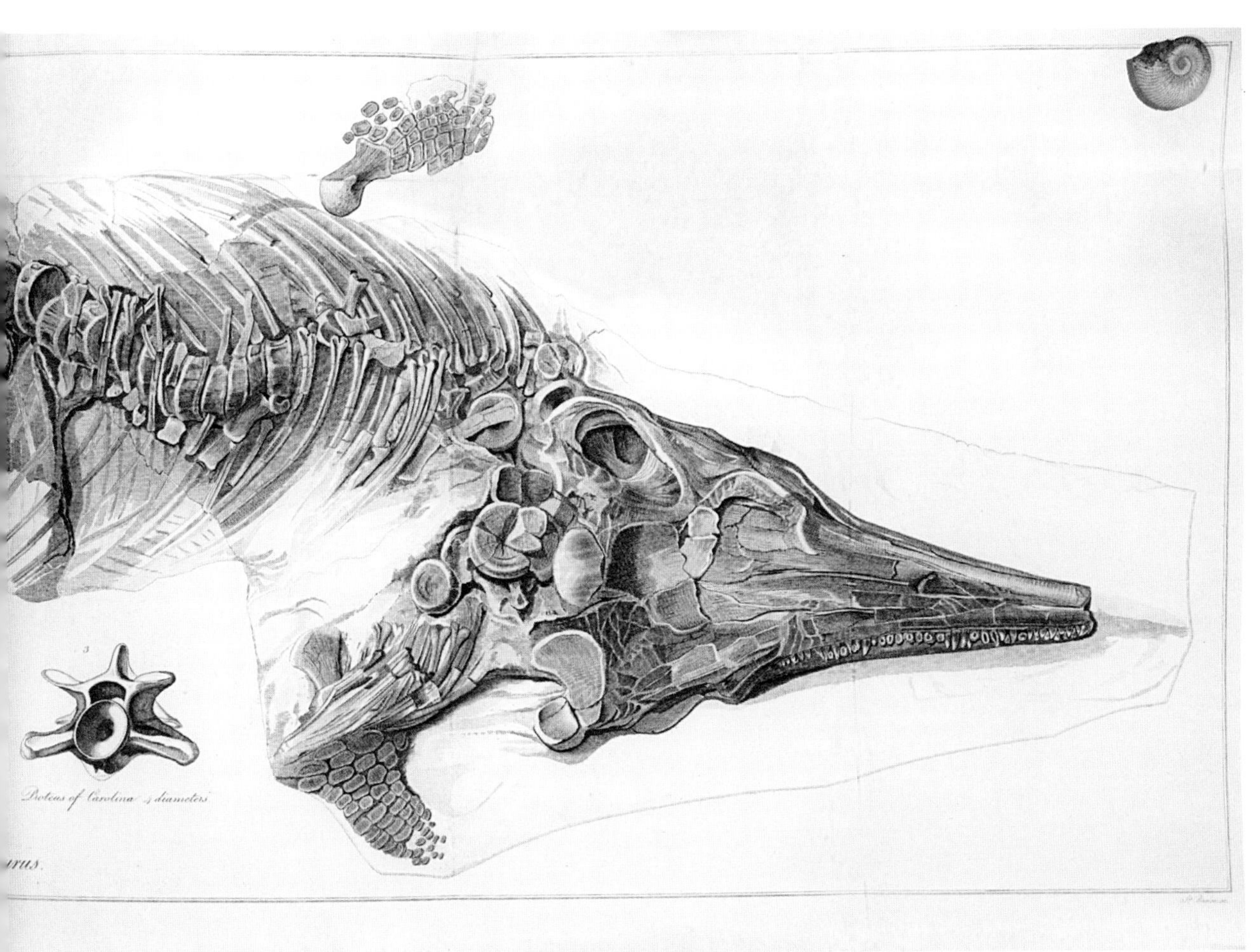

The first scientific illustration of a complete *Ichthyosaur* from 1819.

A DOG SACRIFICED

One of the most colourful figures in the history of European palaeontology is the British fossil expert Mary Anning (1799–1847). In 1811, at the age of 12, she made her first wondrous discovery, the skeleton of an *Ichthyosaurus* in the rocky coast of Lyme Regis in southern England. She also excavated the first *Plesiosaurus* there in 1823. That find was so spectacular that the famous French palaeontologist Georges Cuvier at first believed it to be a forgery at first. He later admitted his mistake. Anning's excavations of marine reptiles from the dinosaur age caused a stir in the upper-class male world of fossil collecting. In 1823, the year of the discovery, *The Bristol Mirror* wrote: 'This persevering female has for years gone daily in search of fossil remains of importance at every tide, for many miles under the hanging cliffs at Lyme, whose fallen masses are her immediate object, as they alone contain these valuable relics of a former world, which must be snatched at the moment of their fall, at the continual risk of being crushed by the half suspended fragments they leave behind, or be left to be destroyed by the returning tide – to her exertions we owe nearly all the fine specimens of Ichthyosauri of the great collections.'

In a letter written in 1833, Anning confessed to a geologist friend that she sometimes risked her life (and that of her dog) during her digs. 'Perhaps you will laugh when I say that the death of my old faithful dog has quite upset me, the cliff that fell upon him and killed him in a moment before my eyes, and close to my feet... it was but a moment between me and the same fate.' As a woman, Anning did not receive the recognition she deserved during her lifetime. Although excluded from joining the prestigious, male-only Geological Society of London, the club made her a posthumous member. The *Ichthyosaurus anningae*, which she had excavated in Dorset, was named after her. That fossil is now in the collection of the Museum of Natural History in Oxford. Other finds can be seen in the Natural History Museum in London. Her (romanticised) life story was the subject of the feature film *Ammonite* (2020), in which Kate Winslet played the role of Mary Anning.

———

Mary Anning, the British fossil collector and palaeontologist,
whose poignant life was fictionalised in the film *Ammonite* (2020).

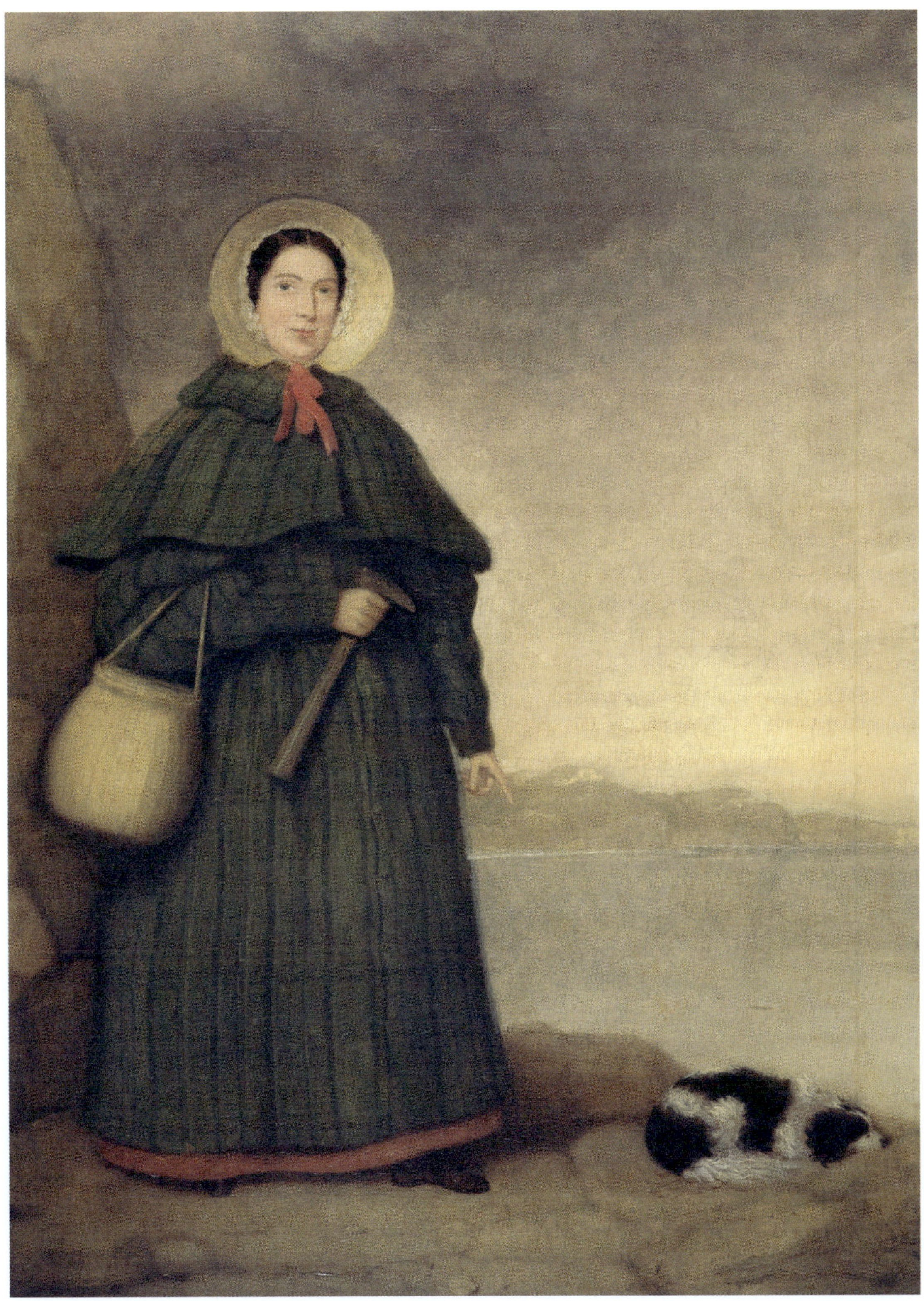

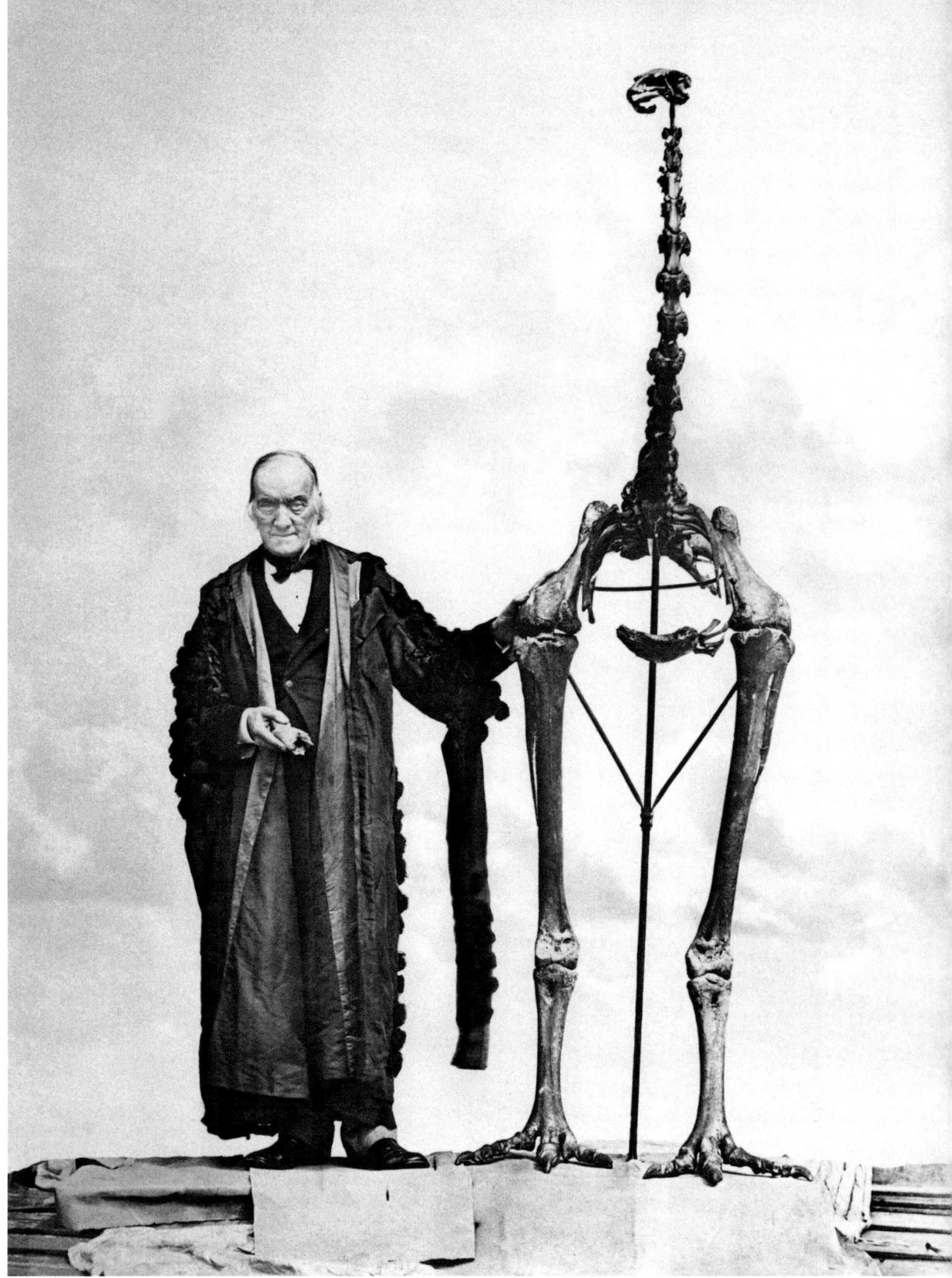

AND THE NAME IS… DINOSAUR

In the early decades of the 19th century, European and American scientists soon became interested in studying the large reptile fossils that began to appear across the world. Englishman Richard Owen (1804–1892) realised that they were not oversized crocodiles or lizards, so in 1841, at a meeting of the British Association for the Advancement of Science, he proposed the name 'Dinosauria'. He wanted to name the 'separate group or suborder of reptiles' and based it on ancient Greek: *sauros* means 'reptile' and *deinos* 'terrible' or 'terrifying'. Owen also immediately recognised that the three genera, *Megalosaurus, Iguanodon* and *Hylaeosaurus*, shared some specific characteristics. This was the main basis for establishing a separate taxonomic group. Owen eventually founded the British Natural History Museum with the support of Prince Albert, Queen Victoria's husband. After political backbiting, the museum finally opened in 1881. Owen was the first person to make the field of the sciences accessible to the general public, which fanned the flames of dinomania.

IGUANA TOOTH

The second dinosaur genus ever described was *Iguanodon*. The finds, a number of teeth, were discovered and illustrated by Mary Ann Mantell (1795–1847). Her husband, Gideon Mantell (1790–1852), described them in a publication in 1822. After further research, he finally named the genus *Iguanodon* (literally 'iguana tooth') in 1825, by analogy with the teeth of the iguana. He couldn't figure out exactly what *Iguanodon* looked like from that tooth, but he estimated that the animal could have been 20m long. 'I was struck with astonishment at the enormous monster which my investigations had called into existence, and was more anxious to reduce its proportions, than to exaggerate them,' he wrote in 1833. Mantell's paper was the second-ever description of a dinosaur. Gideon Mantell and his wife organised further excavations in Tilgate Forest, West Sussex. In 1832, they also officially discovered and described a third dinosaur, *Hylaeosaurus* ('forest reptile'), an animal with body armour belonging to the ankylosaur group.

<

Sir Richard Owen, the British palaeontologist who coined the name 'dinosaur',
poses next to the skeleton of a (now extinct) species of bird
which he named *Dinornis robustus.*

Teeth of the IGUANODON *a newly discovered* FOSSIL ANIMAL, *from the*

Sandstone of TILGATE FOREST, in SUSSEX.

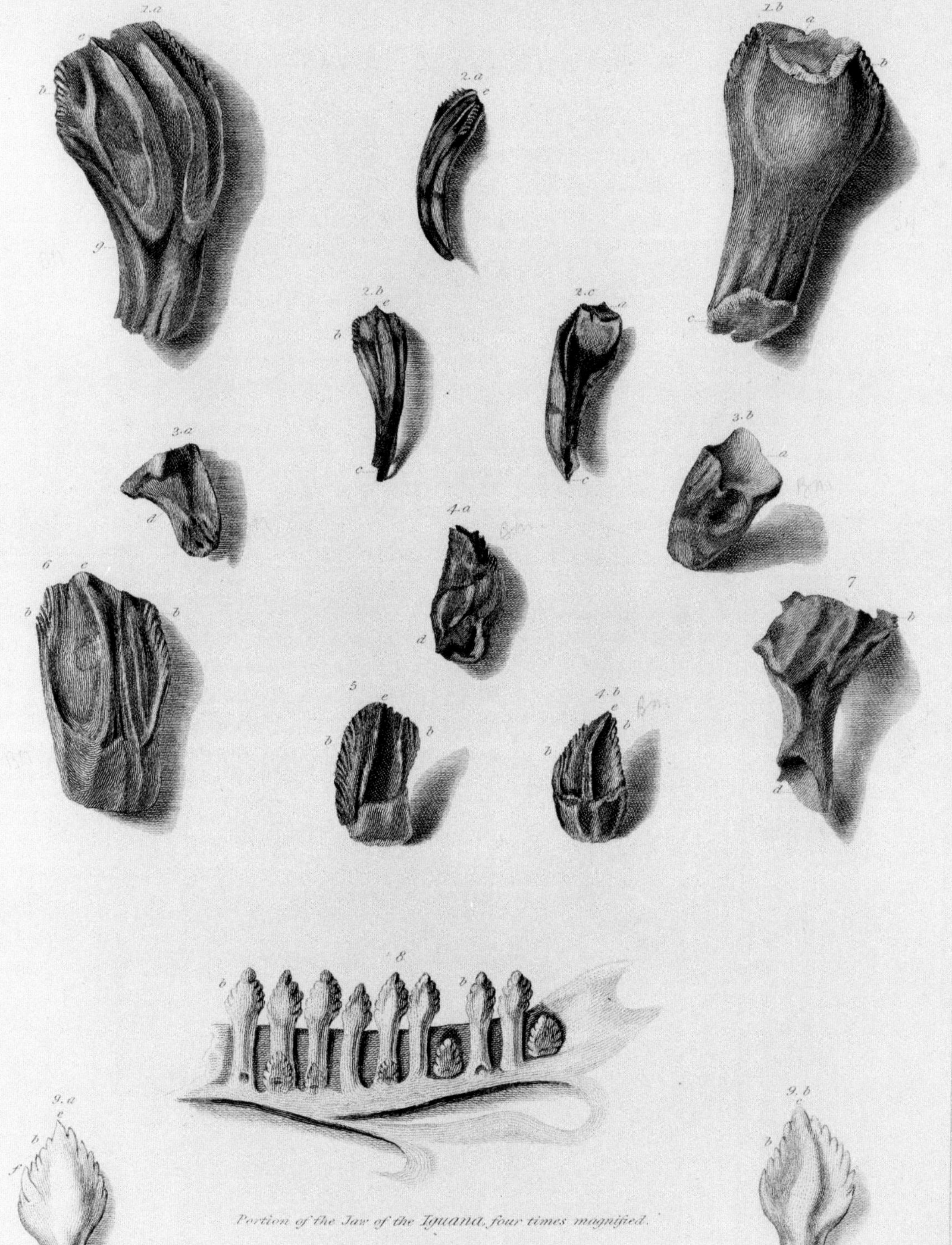

Portion of the Jaw of the Iguana, *four times magnified.*

J.ᵗ Basire sculp.ᵗ

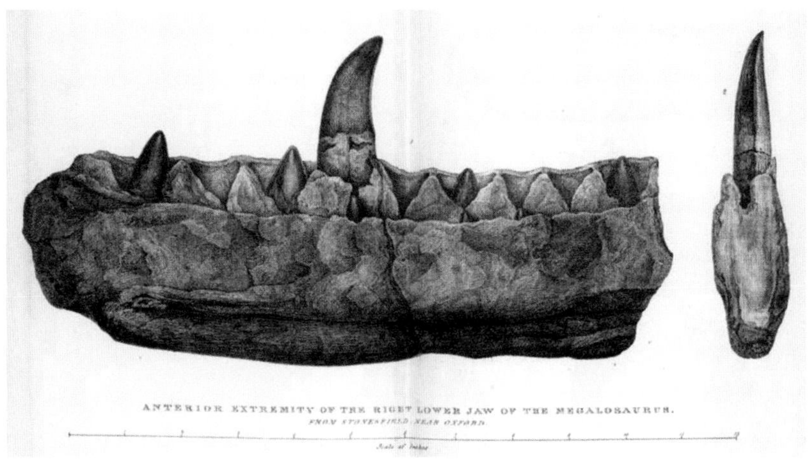

MEGALOMANIACAL SCOOP

The real revolutionary work came from the Reverend William Buckland (1784–1856), who was the very first to describe, illustrate and name a dinosaur in a scientific journal. In 1824, he published the name *Megalosaurus* ('giant reptile') based on bones he had found in Oxfordshire between 1815 and 1818. Although no complete skeleton has been excavated to date, we know that *Megalosaurus* was a theropod dinosaur, 9m long and 4t in weight, that walked on its hind legs. But the first finds were very fragmentary. People drew swift, erroneous conclusions based on these incomplete skeletons. That is why these animals were first depicted as quadrupeds.

∧

The legendary jawbone fossil of a *Megalosaurus*, discovered and described
by the Reverend William Buckland in 1824.

<

Gideon Mantell found these *Iguanodon* teeth in Tilgate Forest in 1822.
Iguanodon was the second dinosaur to be officially described.

THE CRYSTAL PALACE
DINOSAUR THEME PARK

In the mid-19th century, there was still very little scientific research on dinosaurs, and palaeontologists were unable to determine the morphology of dinosaurs accurately. The margin of error was considerable. At the time, some scientists concluded that the *Iguanodon* was a quadruped 60m long, instead of 6–8m. This explains the awkward illustrations of dinosaurs in the scientific publications of the day as well as the almost unrecognisable creatures in the world's first dinosaur theme park, which opened in 1854 at London's Crystal Palace. The gigantic greenhouse that had been built in 1851 for the Great Exhibition in Hyde Park – the first World's Fair – had been bought by a group of investors. They moved the imposing glass building to Sydenham, south London, to run it as a commercial venture. Their plan was to create an educational park around it, depicting the geological and palaeontological history of England. Artist and natural scientist Benjamin Waterhouse Hawkins (1807–1894) was asked to create life-size sculptures of extinct prehistoric animals. Of the 33 colossal sculptures, three were replicas of dinosaurs: *Iguanodon*, *Hylaeosaurus* and *Megalosaurus*. The sculptures were highly speculative because the few fossils offered hardly any information about what they looked like. You have only to look at Richard Owen's drawing of a *Megalosaurus* in his book *Geology and Inhabitants of the Ancient World* from 1854, the year the Crystal Park sculpture garden opened. In it, you can see that Owen deduced the shape of *Megalosaurus* from the few bones that had been found at the time – far too few to reconstruct the animal faithfully. As a result, Owen and Hawkins sculpted a *Megalosaurus* with a hump on its back and four legs – a creature that bears no resemblance to today's portrayals of this bipedal carnivore.

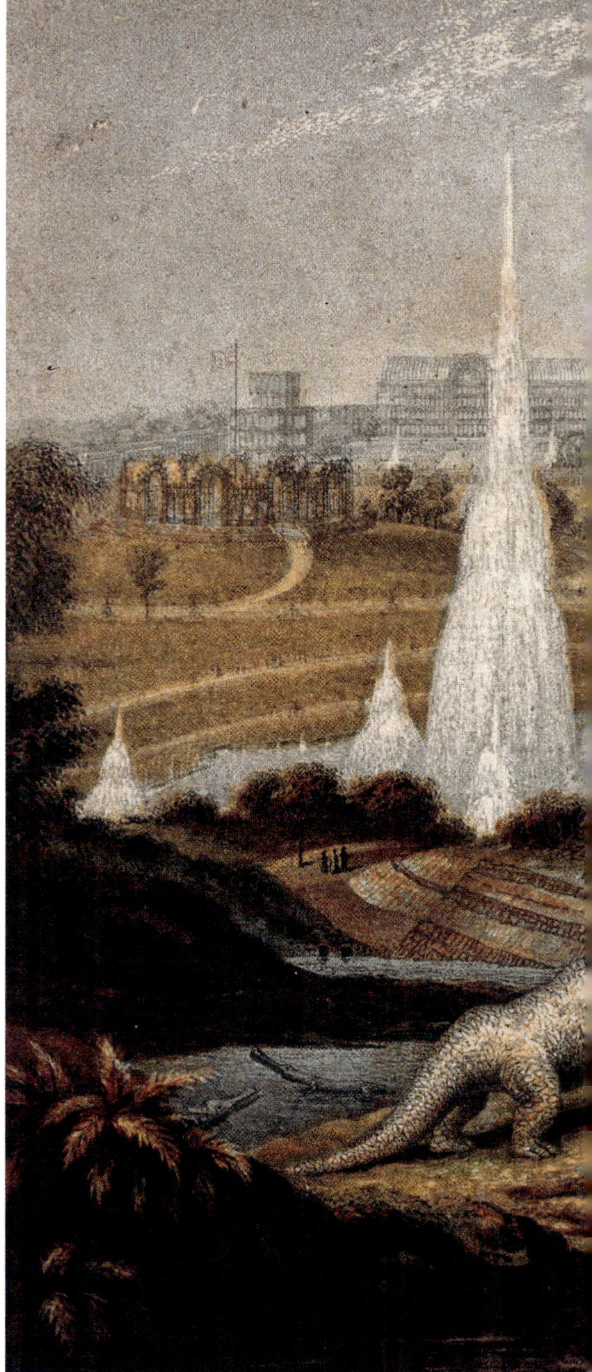

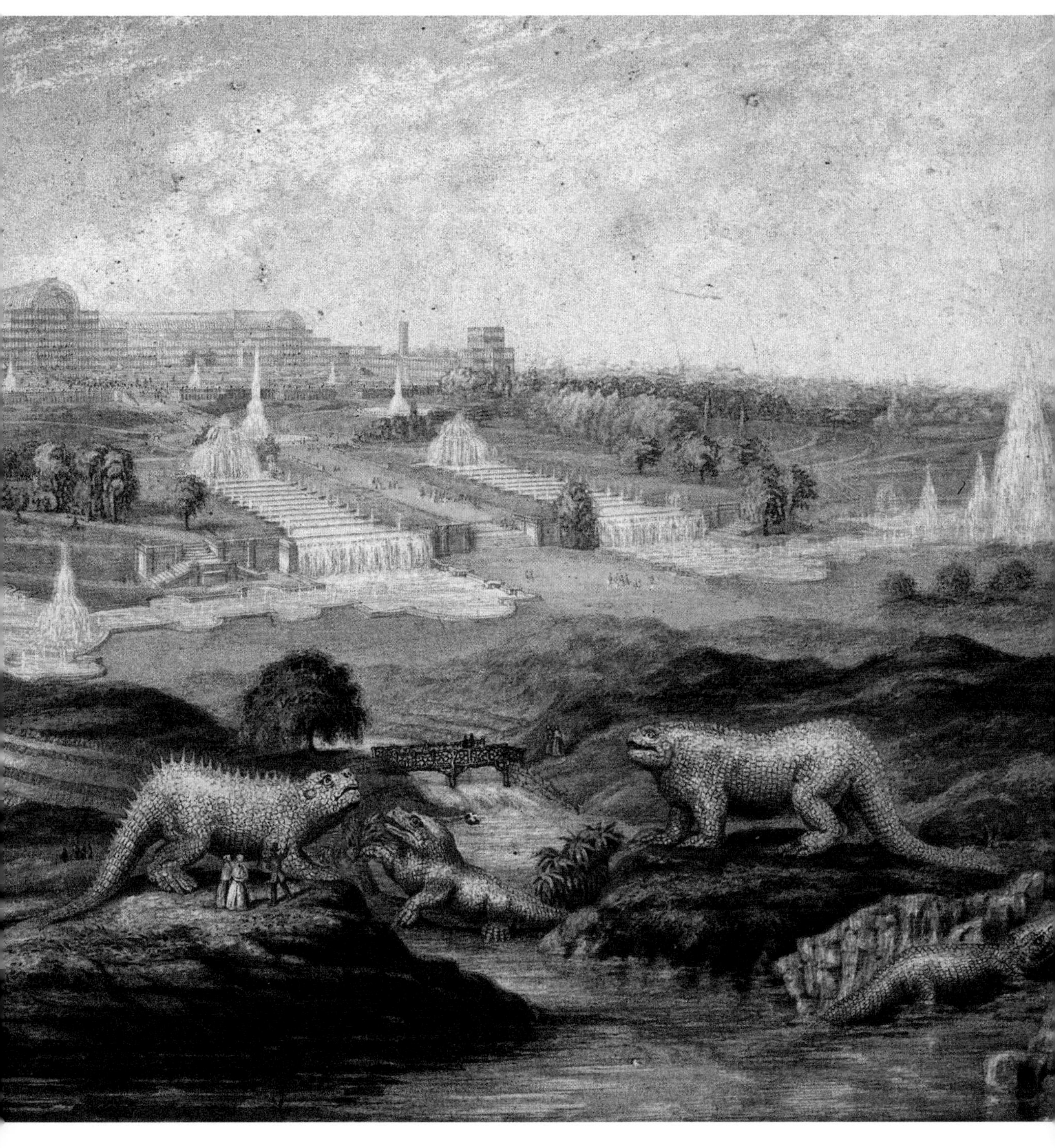

The Crystal Palace from the 1851 World Fair, rebuilt in Sydenham.
It was surrounded by a sculpture park with 'lifelike' sculptures of prehistoric animals.

Despite the conjecture concerning the first dinosaurs' morphology, Hawkins felt scientifically emboldened. He had the open support (and blessing) of Richard Owen, the famous palaeontologist who coined the name 'dinosaur' and tutored Queen Victoria's children in science. Thanks to Owen, he even gained access to the fossil collections of the British Museum, the Royal College of Surgeons and the Geological Society, among others. Just before the prehistoric sculpture park opened in May 1854, Hawkins published a text about it in *The Journal of the Society of Arts*. 'The inevitably fragmentary state of such specimens of course left much to the imagination. It was a fallow field which nothing less than the great enterprise and the resources of the Crystal Palace Company could have attempted for the first time to illustrate and realise – the revivifying of the ancient world – to call up from the abyss of time and from the depths of the earth, those vast forms and gigantic beasts which the Almighty Creator designed with fitness to inhabit and precede us in possession of this part of the earth called Great Britain', he wrote. It is striking that he speaks of the 'Almighty Creator'. Even though Hawkins had created illustrations for Charles Darwin's book *The Zoology of the Voyage of H.M.S. Beagle* (1839–1843), he did not believe in Darwin's theory of evolution, which would be published in 1859.

Hawkins's sculpture park was both a success and a failure. Although many scientists were already questioning the morphology of Hawkins's sculpted dinosaurs, the public were dazzled by his sculptures. Around 40,000 people attended the opening of the open-air exhibition in June 1854, including Queen Victoria. Afterwards, an estimated further two million people visited the concrete prehistoric zoo around the Crystal Palace. Hawkins's park had a lasting influence. With it, he brought fossils out of the dusty, private collections of the elite, where they were seen and studied by a handful of scholars, and allowed the public to enjoy them. At the same time, the park was a business venture, and sold an array of dinosaur merchandise. Within 20 years, the sculptures became – scientifically speaking – hopelessly outdated. But the fantastical concrete statues still stand, tucked away in the shrubbery of Crystal Palace Park. They are a hilarious reminder of just how primitive palaeontology was in Victorian times. Although palaeontologist Michael J. Benton sees it differently: 'We may treat the older images with humour – how could they ever have believed that? – but palaeontologists of the future will likely mock our best efforts.'

———

Benjamin Waterhouse Hawkins had a studio
near the Crystal Palace in Sydenham where he built his dinosaur statues.
He gave a legendary dinner there on New Year's Eve, 1853.

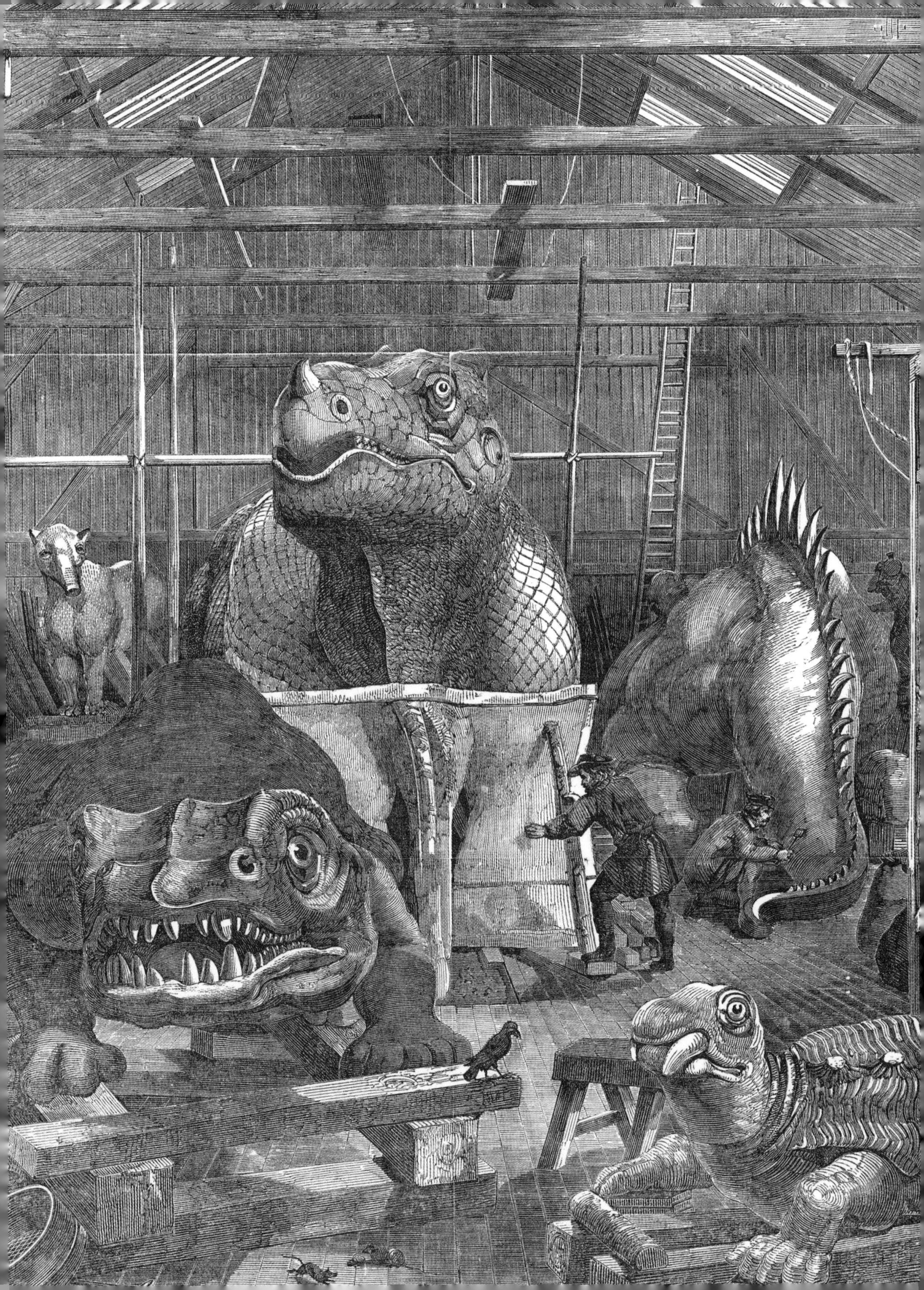

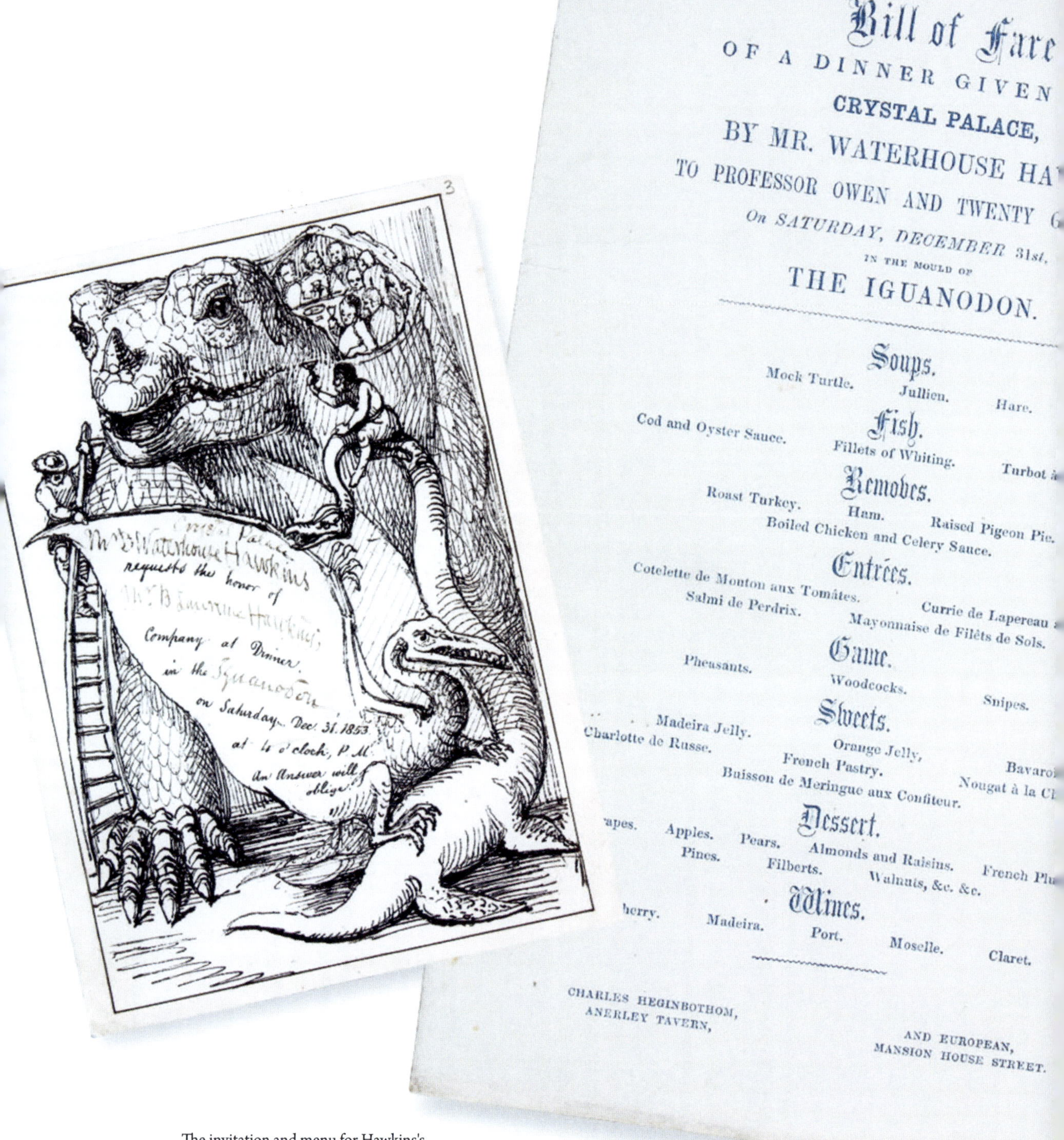

The invitation and menu for Hawkins's eccentric 'dinner in the *Iguanodon*' on New Year's Eve, 1853.

DINNER IN THE *IGUANODON*

Benjamin Waterhouse Hawkins (1807–1894) was PR-savvy and knew how to create a media sensation. In November 1853, he invited Queen Victoria to the studio where he was creating his prehistoric creatures for the sculpture park at the Crystal Palace. That year, the New Year's Eve party in his studio defied all imagination. 'Mr Waterhouse Hawkins requests the honour of – at dinner in the mould of the *Iguanodon* at the Crystal Palace on Saturday evening December the 31st at five o'clock 1853. An answer will oblige,' the invitation read. Hawkins had been preparing his spectacular dinner for over a month. For most of the guests, it was the first time they would see Hawkins's moulds intended for his sculptures for the future Crystal Palace Park. That sculpture park was to show prehistoric animals, including the first described dinosaurs: *Megalosaurus, Iguanodon* and *Hylaeosaurus*. The dinner was held – appropriately – in one of Hawkins's *Iguanodon* models in his Sydenham studio. To keep the 20 guests somewhat warm on that cold winter's evening, Hawkins had erected a sort of circus tent around the iguanodon model, which the guests could enter by a set of stairs. Inside the giant animal, a table was laid out. Symbolically, as the 'brain' of the group and as Hawkins's scientific advisor, palaeontologist Richard Owen sat at the head of the table, literally in the head of the *Iguanodon*. Palaeontologist Edward Forbes, an advisor to Charles Darwin, was also there, as was ornithologist John Gould, known for his illustrated bird books. Naturally, Hawkins also thought about advertising the event: he invited Herbert Ingram, the publisher of *The Illustrated London News*, a newspaper with a circulation of more than 150,000 copies. Around midnight, Edward Forbes stood up to recite a poem:

A thousand ages underground
His skeleton had lain;
But now his body's big and round,
And he's himself again!

Beneath his hide he's got inside
The souls of living men,
Who dare our Saurian now deride
With life in him again?

The jolly old beast
Is not deceased,
There's life in him again.

The PR strategy worked: the dinner is immortalised in a print by Hawkins, which was published in *The Illustrated London News* of 7 January 1854. Herbert Ingram wrote: 'The guests were evidently well pleased with the modern hospitality of the *Iguanodon*, whose ancient sides there is no reason to suppose had ever before been shaken with philosophic mirth.' *The Times of London* talked of 'gigantic tenants of … the earliest and sublimest pages in the history of creation'. Dinomania had begun.

DINOMANIA AND THE FIRST AMERICAN DINOSAUR

The scientific study of dinosaurs really took off in the second half of the 19th century. After the first spectacular discoveries, newspapers and magazines also fell under the spell of dinomania. Those first discoveries and drawings of prehistoric reptiles made everyone dream of immense monsters that roamed the Earth millions of years ago. In 1858, William Parker Foulke (1816–1865) made a milestone discovery; he excavated the first dinosaur on American soil, the *Hadrosaurus*. It was an historic find because it was the first virtually complete skeleton of a dinosaur ever found. The skeleton was described by the palaeontologist Joseph Leidy (1823–1891) as *Hadrosaurus foulkii*. Leidy knew it was a dinosaur because he saw similarities with the *Iguanodon* that had been discovered earlier in England. He was also convinced that *Hadrosaurus* walked on its hind legs. In the *Proceedings of the Academy of Natural Sciences of Philadelphia* in 1858, he wrote: 'The great disproportion of size between the fore and back parts of the skeleton of *Hadrosaurus* leads me to suspect that this great extinct herbivorous lizard may have been in the habit of browsing, sustaining himself, kangaroo-like, in an erect position on its back extremities and tail.' A revolutionary idea because until then most scientists had thought that all dinosaurs were quadrupeds.

The skeleton of *Hadrosaurus* went on display, mounted on two legs, at the Philadelphia Academy of Natural Sciences in 1868. This was also a first, because never before had a dinosaur skeleton been assembled in a museum. The team that set up the exhibition included the British natural scientist Benjamin Waterhouse Hawkins, whose life-size sculptures for Crystal Palace had also made him famous in the United States. For the Philadelphia assembly, Hawkins devised a revolutionary system for attaching the bones to a metal frame so that they approximated the dinosaur's shape. He made plaster reconstructions of the missing legs so that the skeleton looked complete. *Hadrosaurus* – a 9m herbivore – was a huge success. Visitors were awed by the skeleton, which was the talk of the town for months; peopled clamoured to see it. Even though the museum was only open two afternoons a week, around 100,000 people came to see it in 1869: twice the number of visitors the museum had attracted the year before. In fact, the exhibit was such a success that, for the first time in history, the museum management decided to charge an admission fee. However, the enormous crowds did not please everyone at the museum. The annual report of 1869 states: 'The crowds lead to many accidents, the sum total of which amounts to a considerable destruction of property, in the way of broken glass, light wood work, etc. Further, the excessive clouds of dust produced by the moving crowds, rest upon the horizontal cases, obscuring from view their contents, while it penetrates others much to the detriment of parts of the collection.'

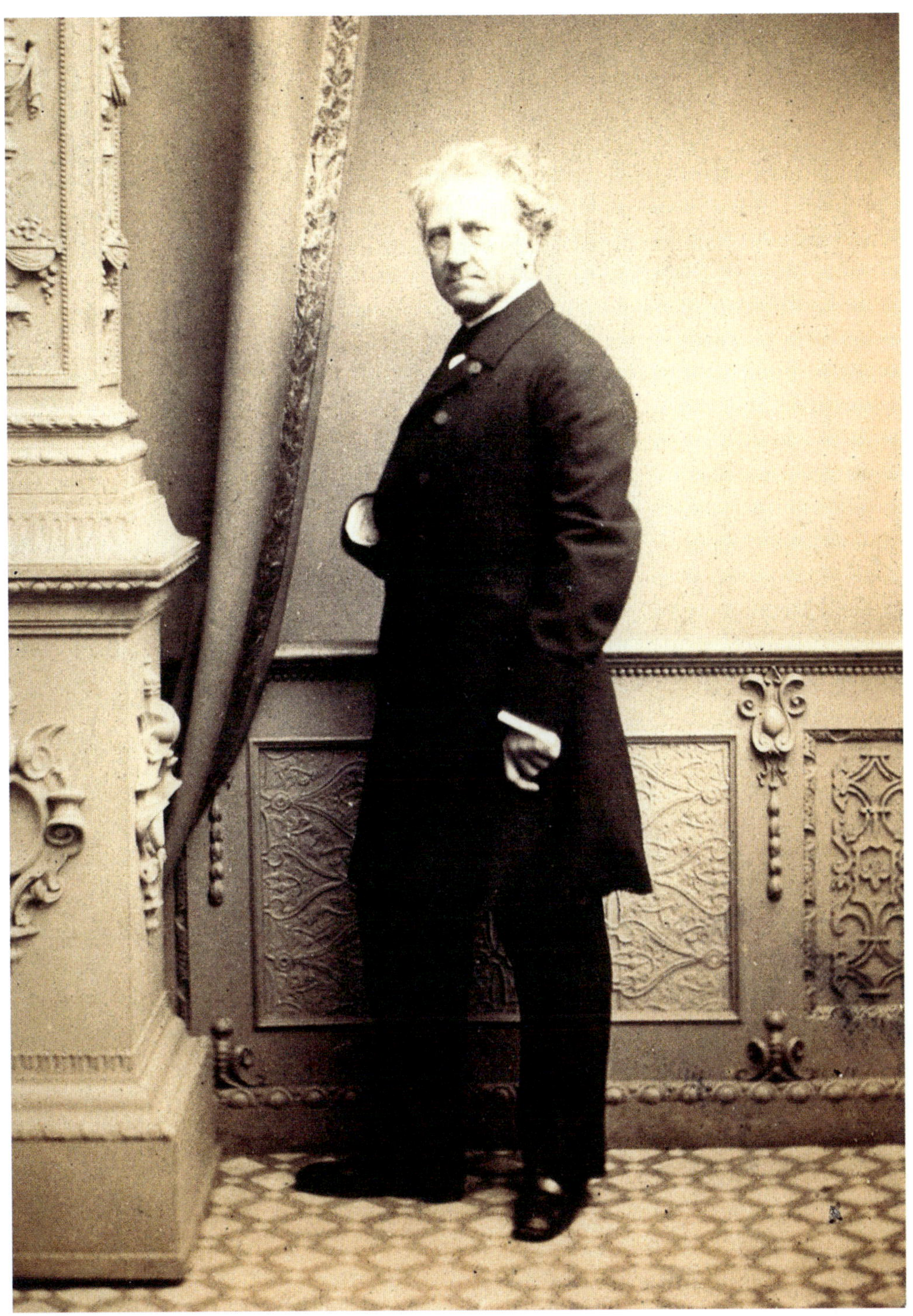

Portrait of Benjamin Waterhouse Hawkins, a Victorian maverick who helped spark dinomania and set up the first dinosaur skeleton in the United States.

DINOSAUR MUSEUM IN CENTRAL PARK

The commercial success of Hawkins's Crystal Palace dinosaur park, and his mounting of the *Hadrosaurus* in Philadelphia did not go unnoticed. In 1868, he was invited to begin a Palaeozoic Museum in New York's newly laid Central Park. The plan was to build a spectacular grouping of dinosaur statues there, similar to those he had created in Sydenham. Seven of those models were built in Hawkins's studio. However, when Hawkins fell out with the corrupt New York politician William M. Tweed, known as 'Boss Tweed', the plug was pulled on the promising museum project in 1870. Worse still, Tweed hired a gang of vandals to smash up Hawkins's studio on 3 May 1871. Nothing remained of the moulds and sculptures Hawkins had been working on for the past three years. And the Dinosaur Museum in Central Park never materialised.

A *DIPLODOCUS* AS PEACE ENVOY

Andrew Carnegie (1835–1919) was one of the richest and most generous entrepreneurs of all time. In the last 20 years of his life, the Scottish American industrialist gave away almost 90 per cent of his immense fortune to education, culture and science. It is to him that we owe the famous Carnegie Hall concert venue in New York, as well as the Carnegie Museums in Pittsburgh and numerous scientific institutions. In order to furnish his newly founded Carnegie Museum of Natural History, the businessman sponsored several palaeontological expeditions. In 1899, one of them was led by John Bell Hatcher, who unearthed the first *Diplodocus* to be discovered in Wyoming. He named the species *Diplodocus carnegii*, after the sponsor of the mission, although the skeleton is now known as 'Dippy'. The excavation of Dippy made headlines worldwide. Because he wanted the whole world to be able to admire the find, Carnegie had several replicas made of the skeleton. He donated them to museums from London to Saint Petersburg, from Madrid to Berlin, from Vienna to Mexico City. It became a kind of prehistoric 'Trojan horse', because for Andrew Carnegie the travelling *Diplodocus* held a hidden message. 'Carnegie hoped to demonstrate through mutual interest in scientific discoveries that nations have more in common than what separates them. He used his gifts in an attempt to open international dialogue on preserving world peace a form of "dinosaur diplomacy",' said William Thomson, Carnegie's great-grandson. Unfortunately, Dippy could not prevent the impending First World War. During the Second World War, the replica in the Natural History Museum in London was even moved to the basement.

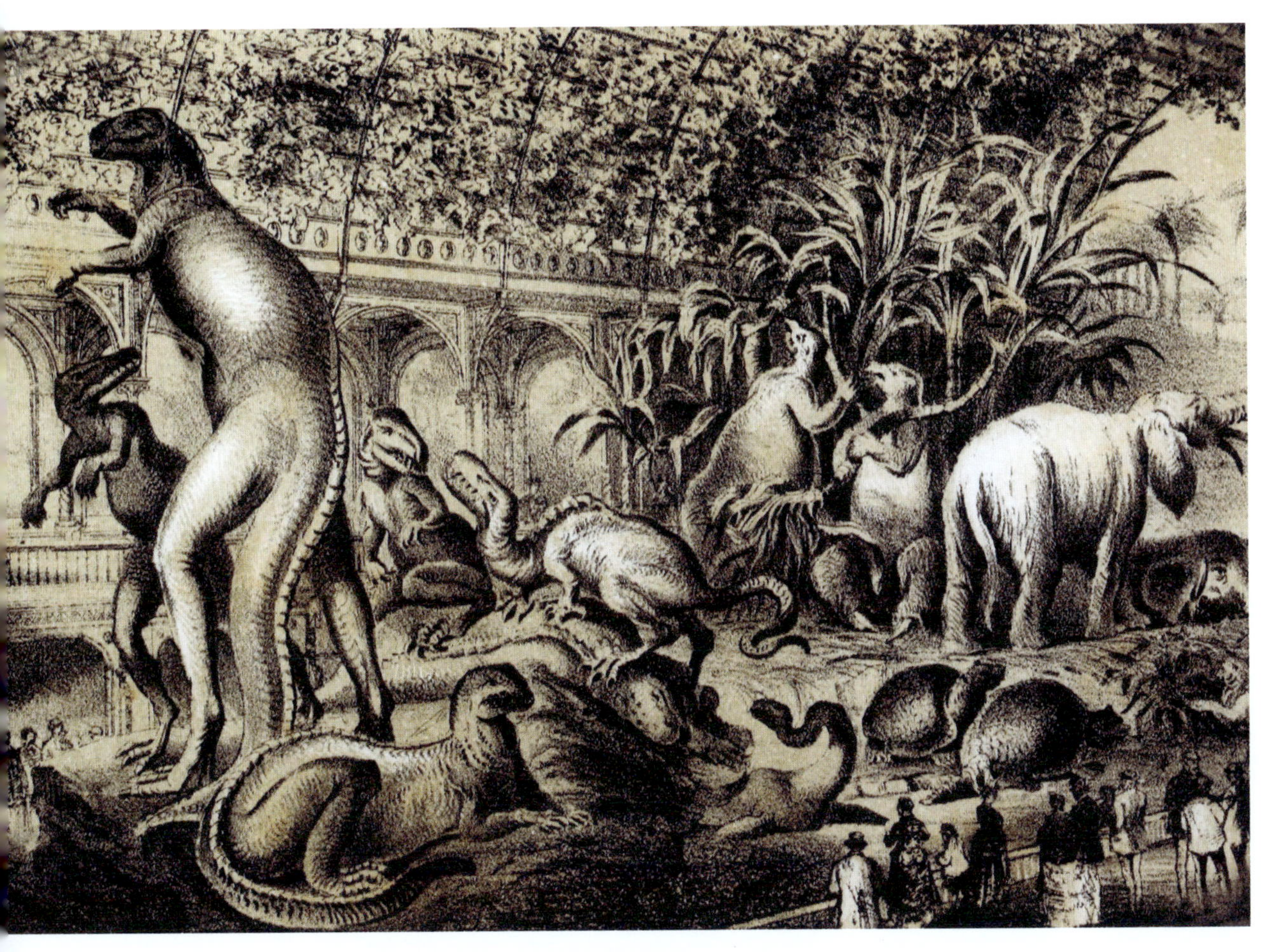

Design drawing for the Paleozoic Museum in New York's Central Park.
Unfortunately, Hawkins's prestige project was never realised,
after hired thugs destroyed his studio in 1871.

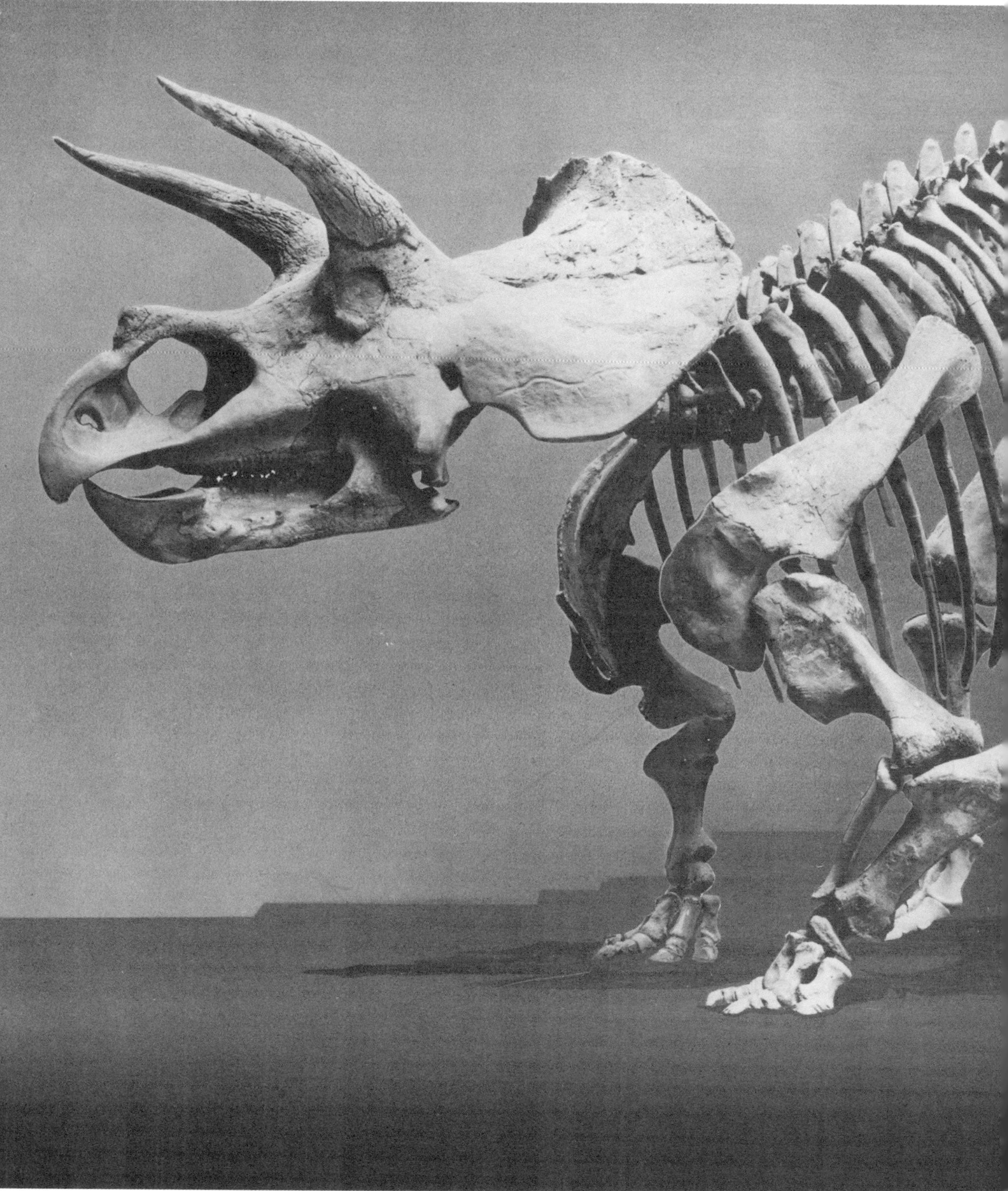

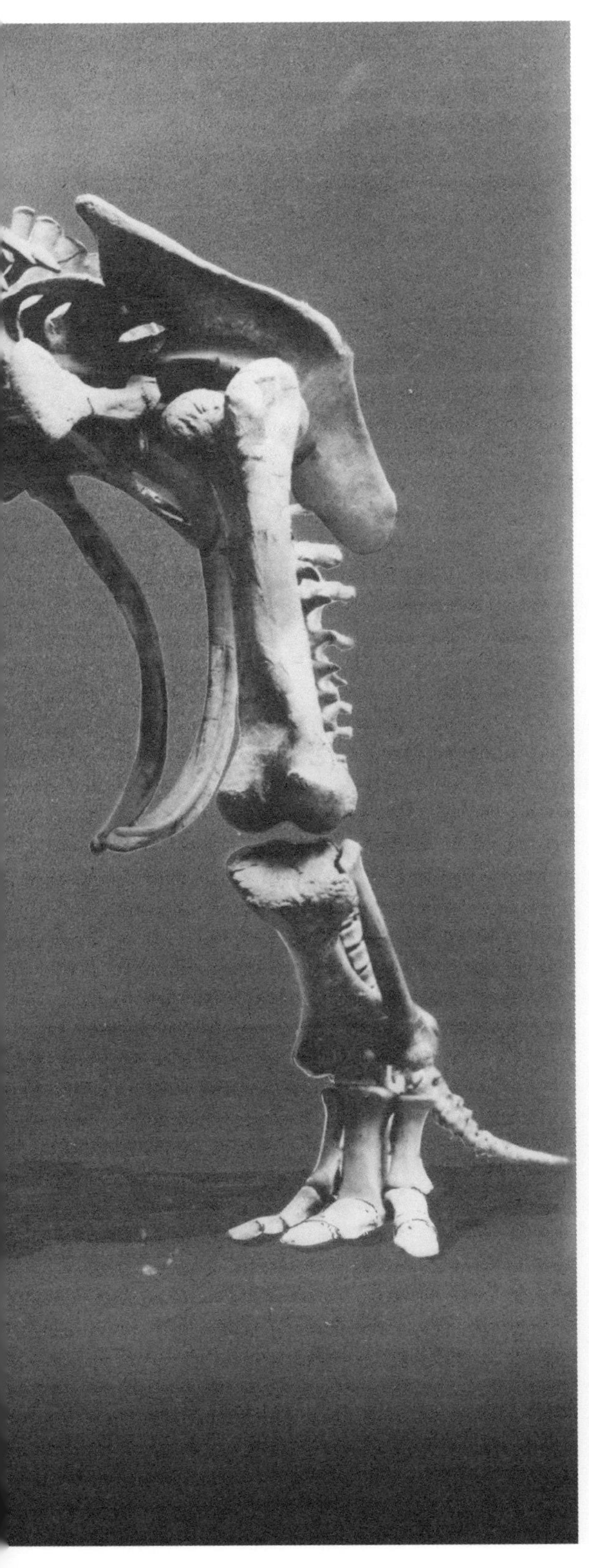

THE RISE AND FALL OF 'HATCHER'

Triceratops is the official dinosaur of the US state of Wyoming. Legend has it that a cowboy, Edmund Wilson, discovered the skull of a huge beast in a Wyoming ravine in 1888. When he lassoed one of the protruding horns, it broke off. Fossil hunter John Bell Hatcher (1861–1904) was able to see the horn and excavate the skull. Because the prehistoric creature had three horns, it was given the name *Triceratops*. For many years, Hatcher worked with Othniel Charles Marsh (1831–1899), the famous scientist and fossil hunter who discovered no fewer than 80 dinosaur species. By 1892, Hatcher had already excavated more than 30 *Triceratops* skulls for Marsh. A year after Hatcher's death from typhoid fever, in 1905, a *Triceratops* was exhibited for the first time ever at the United States National Museum in Washington, now the Smithsonian National Museum of Natural History. This skeleton, nicknamed 'Hatcher', drew crowds and created a huge buzz. 'The most fantastic and grotesque of all that race of giant lizards known as dinosaurs', wrote *The Washington Post* at the time. For 20 years, Hatcher was the only complete *Triceratops* skeleton in the world in a museum. It was the showpiece of the famous 'Hall of Extinct Monsters', but, in actuality, the skeleton looked odd: the creature seemed to have been put together from fragments of a dozen different specimens. The dinosaur's posture and proportions were extremely unnatural and proved such a 'Frankenstein 's monster' that it was completely overhauled in 1998. It was given new casts of other bones, so that in terms of morphology and proportions it looked somewhat 'lifelike'. In 2019, his fate was decided: in the renewed museum set-up, Hatcher was rehomed. Now, he is shown being trampled by a *Tyrannosaurus rex* skeleton, poised to devour the *Triceratops*.

'Hatcher' was the first *Triceratops* to be exhibited ever at the Smithsoninan National Museum of Natural History.

'BONE WARS': WHEN TWO DOGS FIGHT OVER A DINOSAUR BONE...

When William Parker Foulke (1816–1865) discovered the upright-walking *Hadrosaurus* in 1858, it triggered a rush on dinosaur fossils in America. Dinomania reached its peak (or low point, as the case may be) during the so-called 'Bone Wars', a bitter rivalry between two American scientists, Edward Drinker Cope (1840–1897) and Othniel Charles Marsh (1831–1899). The two were originally friends, even naming new dinosaur species after each other. But their contrasting personalities and backgrounds spurred mutual rivalry and envy. And things got seriously out of hand in the 1870s. 'If Cope was lightning, Marsh was thunder,' writes Zoë Lescaze in her book on paleo-art (see Chapter 4). The two palaeontologists engaged in public displays of animosity, verbal and physical. For 30 years, they tried to outdo each other in the search for new dinosaurs (and other fossils). Their most important discoveries were made in Como Bluff, Wyoming. Marsh called Cope 'my bitter enemy'. Palaeontologist Joseph Leidy, once Cope's teacher at the University of Pennsylvania, was appalled by their behaviour: 'I can't stand this fighting. It disgusts me and I am going to drop palaeontology and have nothing more to do with it; because of the way Marsh and Cope are in each other's wool all the time.'

Cope and Marsh were so passionate about fossil discoveries that they resorted to the most brutal methods, stealing each other's finds, and sometimes mixing up the bones of different skeletons. Now and then, they destroyed each other's excavation sites and occasionally attempted to prevent each other's work from being published, simply to have the scientific scoop on a particular find. Cope and Marsh worked with large teams of diggers, who regularly used dynamite to dig up bones, with the sad result that important specimens were irrevocably damaged. The use of dynamite also led to the destruction of many fossils' stratigraphic context: they were no longer in their original layers, which made scientific research far more challenging.

Although such methods are no longer tolerated today, both palaeontologists left an awe-inspiring and unrivalled mark on dinosaur palaeontology. In total, they discovered no fewer than 136 recognised species: Marsh found 80, Cope 56. But Cope wrote five times as many books and scientific papers. However destructive and unorthodox the pair's cowboy mentality was, their 'Bone Wars' eventually led to the discovery of some well-known dinosaurs, such as *Allosaurus, Stegosaurus, Brontosaurus, Triceratops* and *Apatosaurus*. Palaeontologist Robert Bakker summed it up nicely: 'The dinosaurs that came out of Como Bluff not only filled museums, but also magazine articles, textbooks and everyone's mind.'

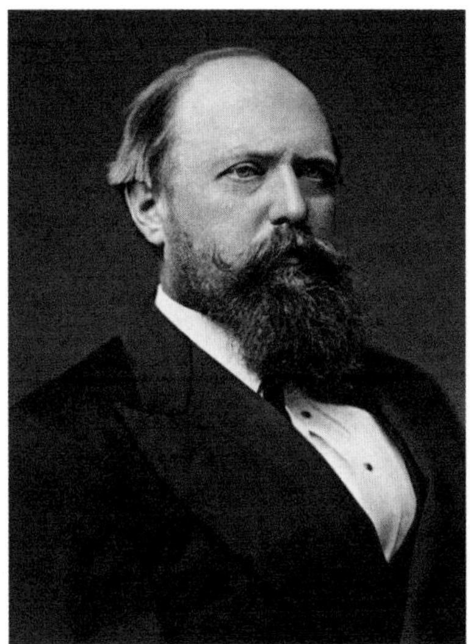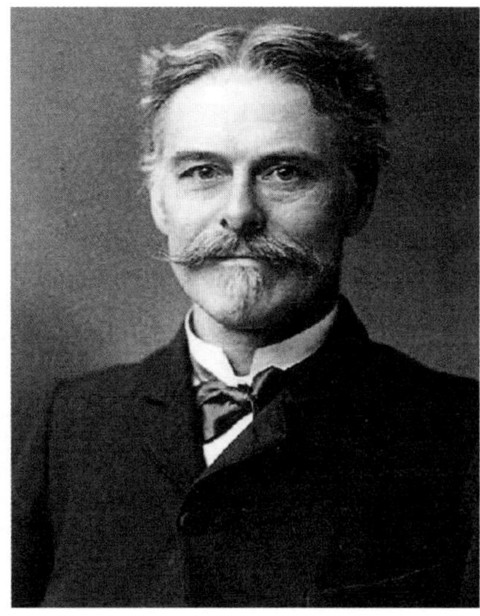

The American palaeontologists Othniel Charles Marsh (left) and Edward Drinker Cope (right) fought the 'Bone Wars' at the end of the 19th century: an uncollegiate pursuit of dinosaur remains and publications.

The 'Bone Wars' lasted 30 years. By the time they died, the two men had invested such huge sums of money in their palaeontological rivalry that they were both impoverished. But even after their deaths, Cope would not bury the hatchet. He had arranged for his remains not to be buried but donated to science. He challenged Marsh to do the same, so that, after their deaths, researchers could measure who had the largest brain. At the time, brain size was believed to be a way of determining intelligence. Cope's brain weighed 1.54kg. How heavy Marsh's was is unknown, as he did not accept Cope's challenge. Cope's brain is preserved at the Wistar Institute and his skull is in the collection of the University of Pennsylvania Museum of Archaeology and Anthropology. Some sources claim that Cope also tried to have his own skeleton recorded as type material for the species *Homo sapiens*, but this was refuted by the Academy of Natural Sciences. Cope's fossil collection is held in the American Museum of Natural History, Marsh's in the Yale Peabody Museum of Natural History. Michael Crichton, the author of *Jurassic Park*, immortalised the story of the two rivals in the book *Dragon Teeth*, which was published posthumously in 2017.

IGUANODONS IN A COAL MINE

While the 'Bone Wars' were raging in North America, a sensational discovery also came to light in a coal mine at Bernissart in Belgium. The unexpected find would put the newly established kingdom on the map in terms of dinosaur palaeontology. In 1878, two miners, Jules Créteur and Alphonse Blanchard, came across a clay subsidence in a coal seam at a depth of 322m. At first, Créteur and Blanchard believed they had found tree trunks and gold in the clay. They were misled by the ubiquitous pyrite, the glimmering, sulphur-rich iron mineral. After inspection, the engineers and geologists realised that they were not looking at tree trunks, but bones. Almost immediately, they realised the significance of their discovery.

They sent some of the remains to Pierre-Joseph van Beneden, Professor of Zoology and Palaeontology at the Katholieke Universiteit Leuven. He recognised the fossils as the teeth of an *Iguanodon*, the species described more than 50 years earlier by Mary Ann and Gideon Mantell. A legendary telegram sent by the mine director on 12 April 1878 has been preserved: 'Important discovery in the Bernissart coal mine. Pyrite is causing the finds to decompose. Send De Pauw tomorrow. We will pick him up at 8am at Mons station.' Louis De Pauw (1844–1918) was the right man in the right place: the head preparer of the Royal Belgian Institute of Natural Sciences had extensive experience in preparing and assembling fossils. However spectacular De Pauw found 'the mine walls full of fossil plants and fish', he and the team of palaeontologists were dogged by bad luck from the start. When they removed the first dinosaur leg, the bones deteriorated rapidly. This was due to contact with the outside air and the influence of the pyrite. A chemical process known as 'pyrite disease' caused the bones to split and crumble. Fortunately, De Pauw developed new techniques to pack the bones, a method that palaeontologists are still using almost a century and a quarter later. Exposed fossil bones were covered with wet paper and a layer of plaster. Once hardened, the entire thing was hacked loose from the clay layer containing the bones. Then the bottom was also covered with paper and plaster. Fortunately, De Pauw made detailed sketches recording the position of the bones on site, and the packed blocks were carefully labelled to identify which fragments belonged to which dinosaur.

>

Gustave Lavalette sketched the fossils of iguanodons and a crocodile
as they were discovered in 1878 in the Bernissart coal mines.

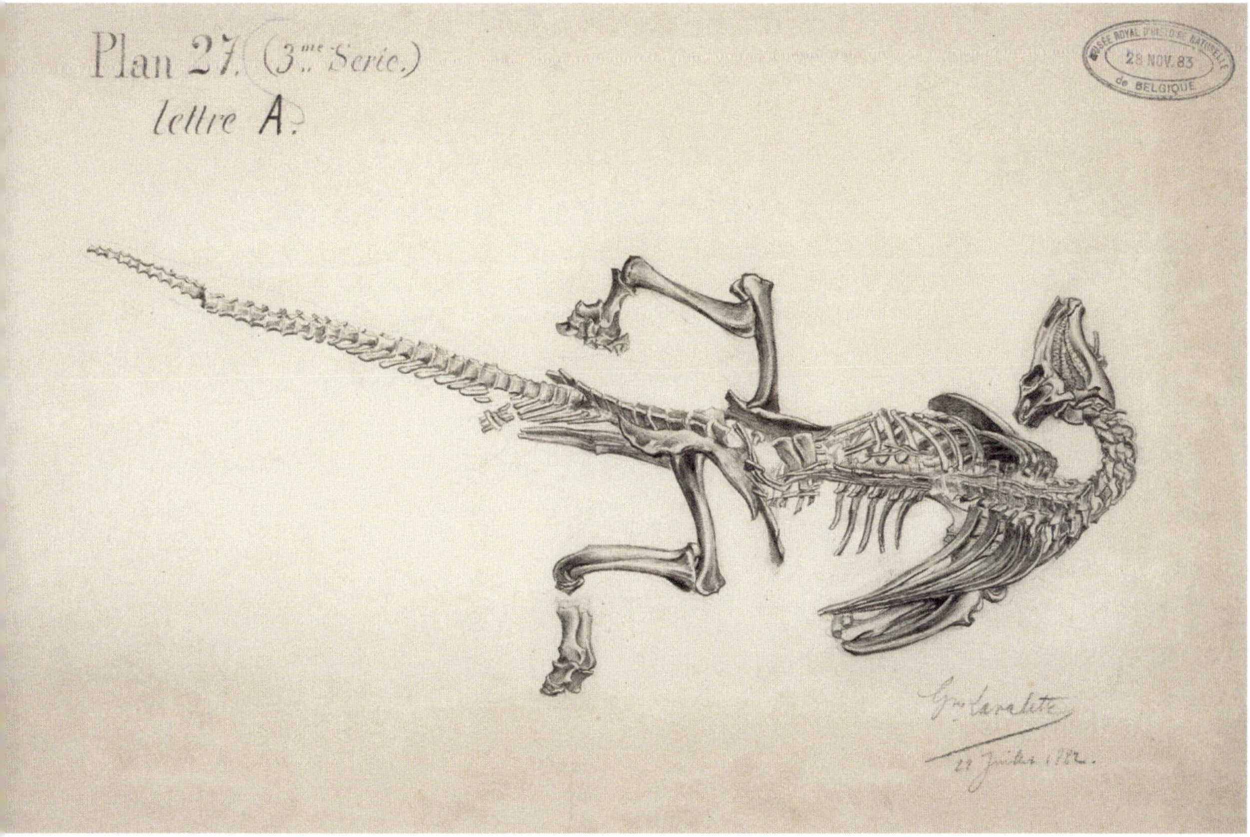

Plan 27. (3me Serie.)
lettre A.

G. Lavalette
29 Juillet 1882.

Bruxelles le 14 9bre 1882
G. Lavalette

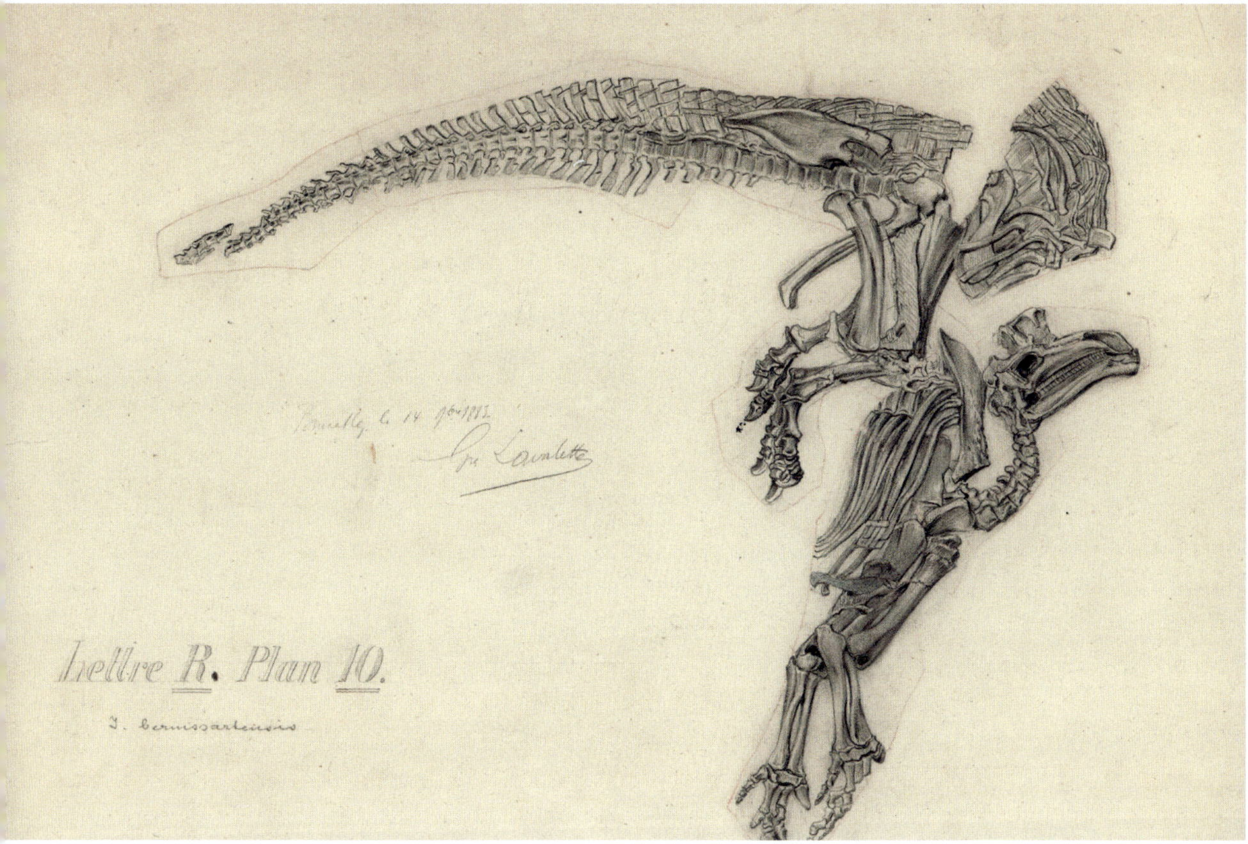

lettre R. Plan 10.

J. bernissartensis

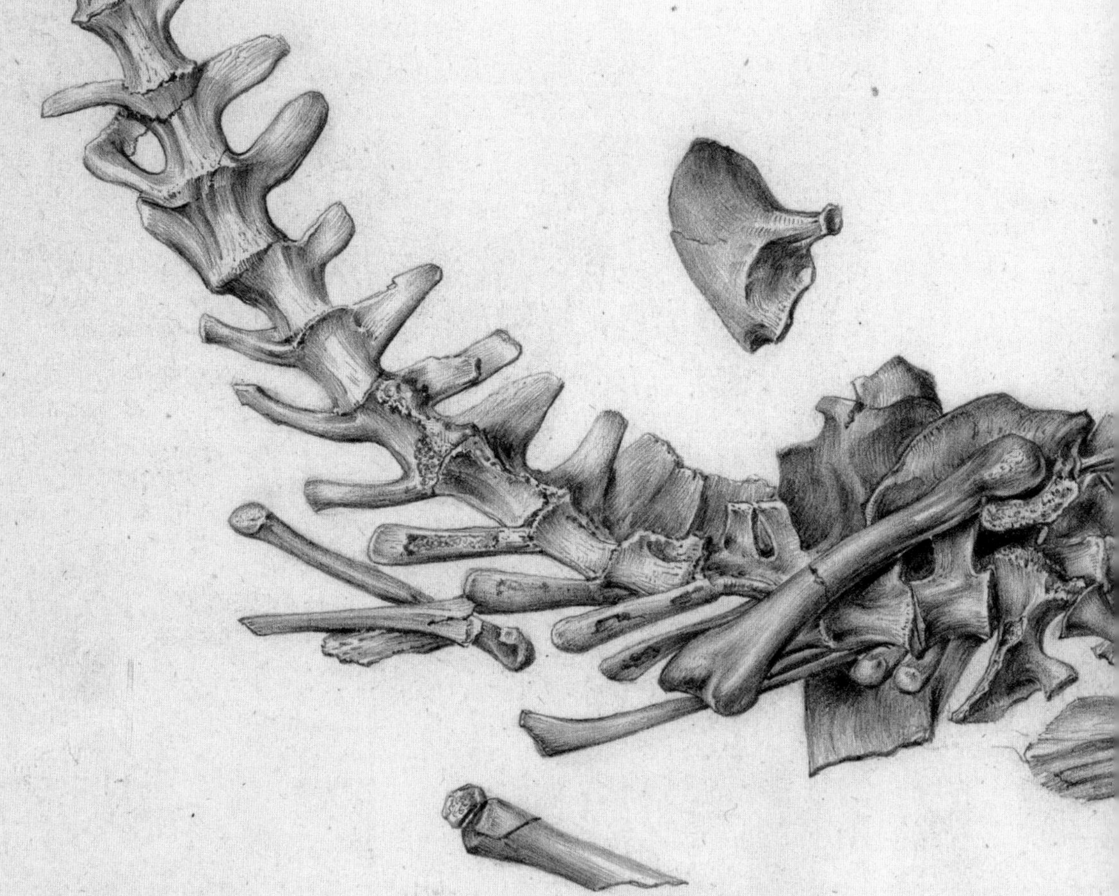
Plan
Coupe

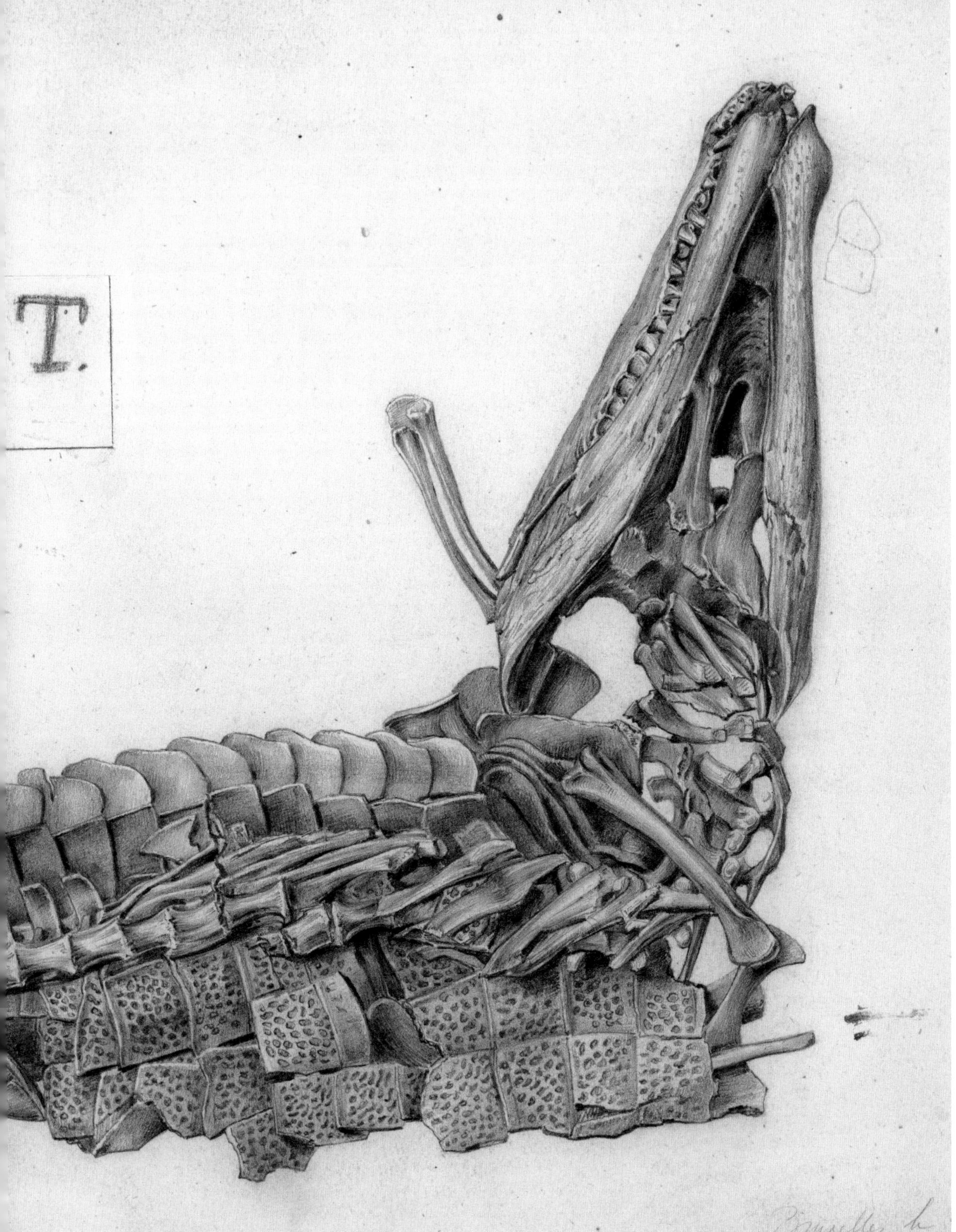

T.

Upon arrival at the workshops of the Royal Belgian Institute of Natural Sciences in Brussels, the fossils underwent further preparation. Essentially, they were treated with gelatin glue to protect them from pyrite decay. Thanks to the detailed excavation notes, artist Gustave Lavalette was able to make accurate in situ drawings of the excavated skeletons. He made precise sketches that clearly showed how the fossils were found in the mine. At a depth of 322 and 356m, these were perilous undertakings, with many risks of collapse, subsidence and flooding. Scientifically speaking, it is still one of the most important dinosaur finds in history. In total, some 30 skeletons were unearthed. Thanks to the swift intervention of De Pauw and the museum, the process was conducted carefully, preserving key scientific information. As a result, scientists today can still extract new data from the archives and the fossils from the Bernissart sinkhole. The *Iguanadon* skeletons and other finds from the Walloon coal mine are still the showpieces of the Museum of Natural Sciences in Brussels – although there is also one *Iguanodon* in the local Bernissart dinosaur museum.

ARCHAEOPTERYX: THE MISSING LINK BETWEEN BIRD AND REPTILE?

Amid the discoveries of huge, awe-inspiring dinosaurs, a pivotal palaeontological find was unearthed in a Solnhofen limestone quarry in southern Germany. In 1861, the first fossil of *Archaeopteryx lithographica* was described by the German palaeontologist Christian Erich Hermann von Meyer (1801–1869). He initially found only a feather, which is why the animal was given the genus name *Archaeopteryx,* which literally means 'ancient feather'. The species name refers to the lithographic limestone in which the fossil was found. The fossil was in the hands of a local doctor, Karl Häberlein, who wanted to sell it for a considerable sum. After lengthy negotiations, the find was eventually bought by the Natural History Museum in London, which was prepared to pay £700 for Häberlein's complete fossil collection – a small fortune, because today that would be the equivalent of £86,000; palaeontologist Richard Owen had strongly urged the museum trustees to acquire the fossil. The Museum für Naturkunde in Berlin also has a famous specimen, found in 1875, which is even better preserved than the one in London.

>

Artist and amateur scientist Léon Becker painted the reconstruction
of an *Iguanodon* skeleton in Brussels in 1884 (see also p. 123). This work of art now hangs
in the Royal Institute of Natural Sciences in Brussels.

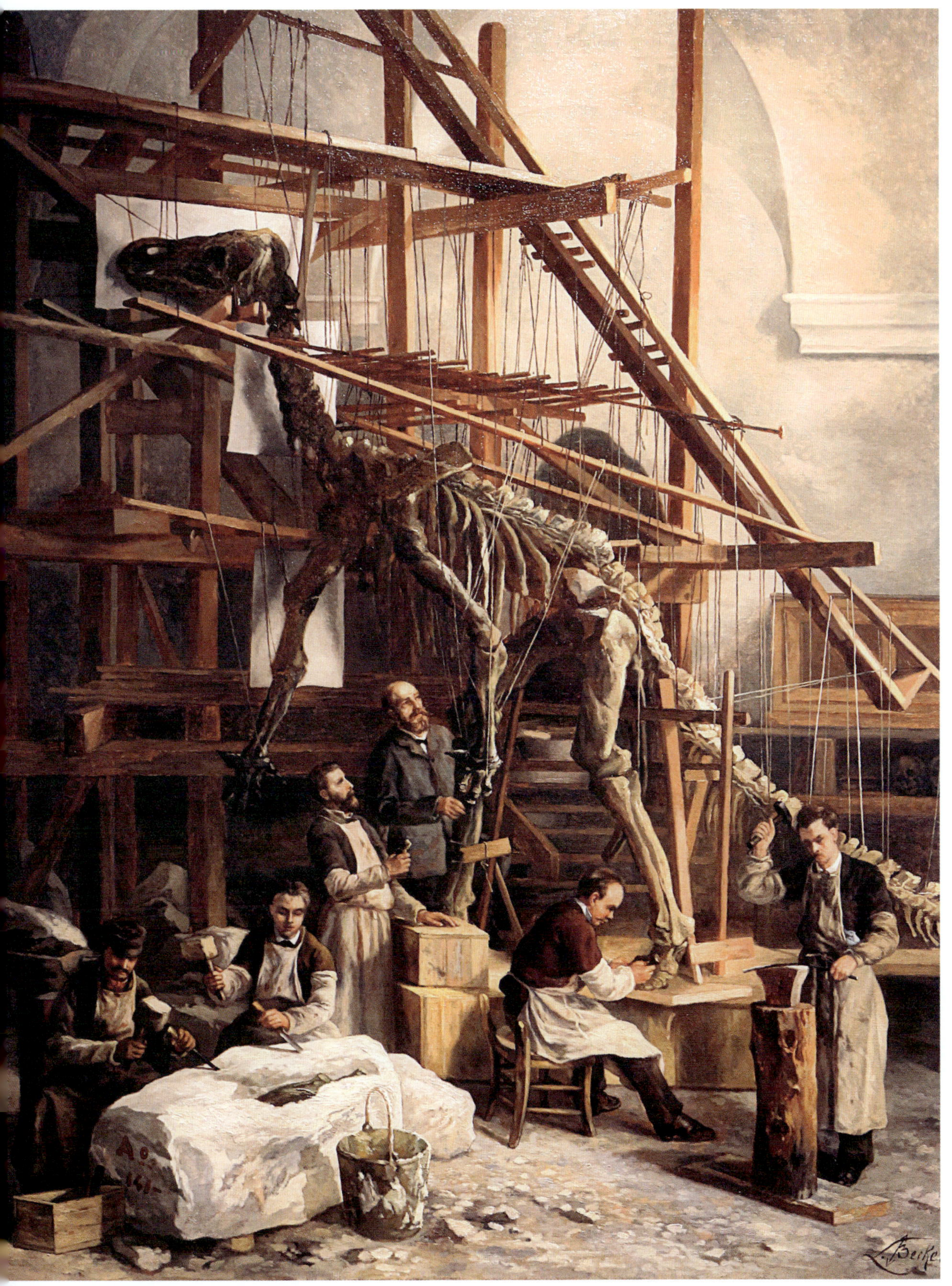

It was soon clear that *Archaeopteryx* was somewhere between a reptile and a bird. It had a long tail, claws, a mouth full of teeth, and a body that had clearly been covered with feathers. The timing of the discovery could not have been better. Charles Darwin had just published his *On the Origin of Species* in 1859, and this discovery instantly became a vital link that confirmed his theory. *Archaeopteryx* was the first dinosaur to be recognised as a bird, although today scientists no longer regard it as a direct ancestor of all modern birds, rather as a side branch in their evolution. Many new finds of early birds, mainly in China between 1990 and 2010, have mapped that family tree and the transition from dinosaurs to birds much better. Today, there are 13 known specimens of *Archaeopteryx*, but the lithographic limestone in Bavaria where it was first found is still being quarried, so there is a good chance that, sooner or later, further discoveries will be unearthed.

MISTER BONES AND THE FIRST *TYRANNOSAURUS*

The 'Bone Wars' between Edward Drinker Cope and Othniel Charles Marsh led to a bona fide American dinosaur rush, which produced important discoveries in the early 20th century. There were countless dinosaur expeditions to the western United States and Canada. Those scientific trips were organised by new, wealthy museums. Just think of the Royal Ontario Museum in Toronto, the Carnegie Museum of Natural History in Pittsburgh, the United States National Museum in Washington, DC (now the Smithsonian Institution), the Field Museum of Natural History in Chicago and the world-famous American Museum of Natural History in New York, which was headed by the palaeontologist Henry Fairfield Osborn (1857–1935). In 1905, Osborn described a monstrous carnivore measuring 12m and weighing 8t, a species which is still hugely popular with the general public: *Tyrannosaurus rex*. The first finds, a pair of teeth, were unearthed in 1874 near Golden, Colorado, by Arthur Lakes, a geologist who later went on to work for Marsh. In 1892, Cope also described a pair of *Tyrannosaurus* vertebrae, which he called *Manospondylus gigas* and wrongly attributed to a horned dinosaur. The first partial *T. rex* skeleton was found in 1900 in eastern Wyoming by Barnum Brown (1873–1963), nicknamed 'Mr Bones'. The flamboyant Brown – he sometimes wore a fur coat while digging – used controlled blasts of dynamite to remove the rocks in the Hell Creek Formation, and found a second partial skeleton in 1902. The first skeleton was originally named *Dynamosaurus imperiosus* by Osborn, but a year later he realised that the two skeletons must belong to the same species. He preferred the name *Tyrannosaurus rex*, which literally means 'King of the Tyrant Lizards'. The first publication on the *T. rex* dates from 1905. In 1908, Barnum Brown found another *Tyrannosaurus* in the same area. Supplemented by casts from his first *T. rex*, the American Museum of Natural History was able to exhibit an almost complete skeleton from 1915 onwards. This spectacular montage inspired the logo and poster image for the film *Jurassic Park*.

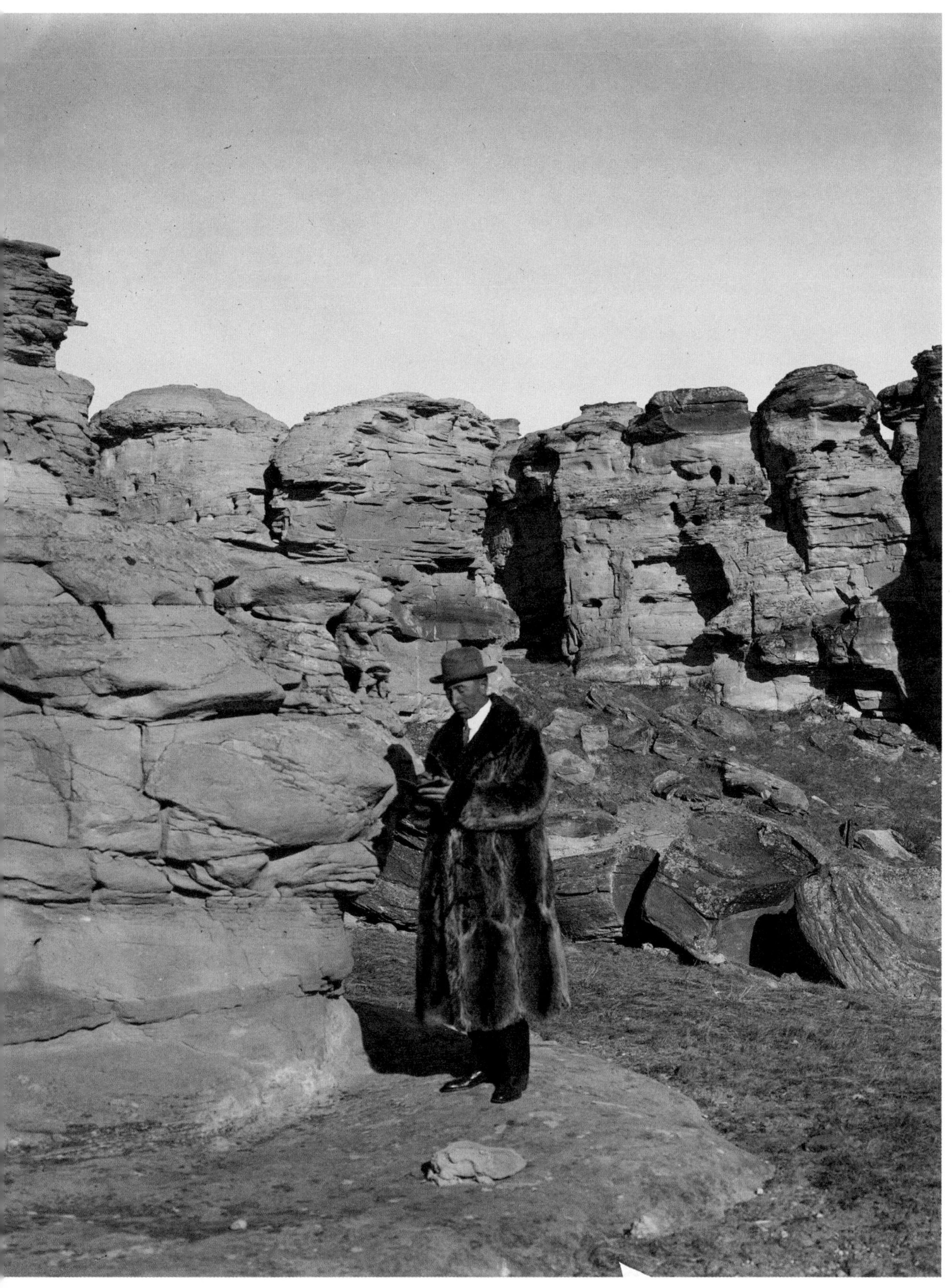

DIGGING IN THE COLONIES

Colonialism is also woven into dinomania. At the beginning of the 20th century, numerous Western nations had colonies and outposts scattered around the world. Dinosaur-hunting expeditions also took place in such places, driven by imperialist motives. Colonialism inspired scientific digs in countries such as Tanzania, Egypt and Mongolia. From 1909 to 1913, the Museum für Naturkunde in Berlin carried out an expedition to Tendaguru Hill in the south-east of present-day Tanzania, which had been part of the German colony of East Africa since 1891. The Germans initially hoped to extract valuable minerals or ores from the soil. Around 1907, a local miner tipped them off about other treasures, 150-million-year-old fossils, prompting the Berlin Museum of Natural History to launch a palaeontological expedition in the former colony.

Within five years, this mission, led by palaeontologist Werner Janensch (1878–1969), yielded around 250t of fossil material for the Berlin museum. The finds arrived in about 800 crates. Every Monday, porters carrying 50–60 boxes walked from the site in Tendaguru to the boat in Lindi, a journey that lasted almost four days. The fossils were packed as carefully as possible in cloth, plaster or clay and transported in bamboo boxes. Photographs of these hikes and of the Tanzanians who carried out the excavations made the national press in Germany. The stories were used as colonial propaganda, for the finds were described by the Germans as 'National treasures on German soil'. Sometimes up to 500 people, both men and women, worked at the excavation sites. These people were paid for their work, and their wages were the largest expense of the entire expedition. The Tanzanian mission yielded a wealth of scientific information. The showpiece was the skeleton of *Giraffatitan brancai* (originally described as *Brachiosaurus brancai*), a long-necked sauropod that lived between 145 and 150 million years ago. The enormous creature still steals the show in the main hall of the museum. To this day, it is the largest assembled dinosaur skeleton in the world, even though it is composed of several specimens of the same species excavated at different sites. The German expedition is still considered one of the most successful-ever excavations. Scientists continue to study finds unearthed during the expedition even today. Due to time constraints, some finds have never even been unpacked.

————

>

Fossil of a 'mummified' *Edmontosaurus*, including a piece of skin.

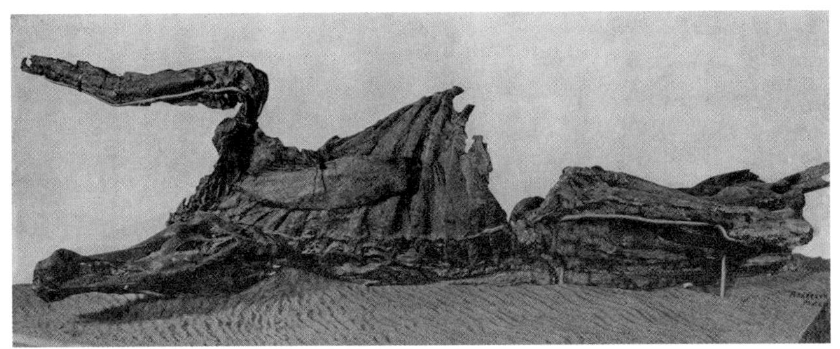

A MUMMIFIED DINOSAUR

In 1908, an extraordinary 'dinosaur mummy' was unearthed in Wyoming. Fossil hunter Charles H. Sternberg (1850–1943) and his three sons came across the almost intact skeleton of an *Edmontosaurus* in Wyoming's Hell Creek Formation in 1908. What was unusual was that the skin had also been largely preserved. Sternberg sold the petrified mummy to the American Museum of Natural History, where numerous other fossils collected by him and his sons are on display. Towards the end of his life, Sternberg often came to the museum to admire his finds. A deeply religious man, he wrote these poetic words after a visit to his famous mummified dinosaur: 'My own body will crumble in dust, my soul return to the God who gave it, but the works of His hands, those animals of other days, will give joy and pleasure to generations yet unborn.'

SANDSTORM AND HARDSHIP

Around the same time, in 1910–1911, Ernst Freiherr Stromer von Reichenbach (1871–1952) organised an expedition to Egypt on a hunt for the world's oldest mammals. At the time, Egypt was a British colony and the mounting tensions between Germany and Britain caused some difficulties in obtaining a permit to explore the desert. Stromer was eventually granted a permit, although the expedition was a disaster. The team had to contend with sandstorms and an extremely harsh winter. But in the end, Stromer did not come across layers from the Eocene period, but from the Cretaceous – tens of millions of years older. The fossils he brought back included the giant crocodile *Stomatosuchus*, the dinosaurs *Aegyptosaurus*, *Bahariasaurus* and *Carcharodontosaurus*, and the remarkable *Spinosaurus aegyptiacus*. Unfortunately, most of his collection was lost in 1944. During a raid by the Allies at the end of the Second World War, the museum in Munich, where the only *Spinosaurus* and *Aegyptosaurus* skeletons were kept, was heavily bombed.

THE INDIANA JONES OF PALAEONTOLOGY

Between 1920 and 1928, the American Museum of Natural History organised a number of expeditions to Mongolia and other parts of Asia under the leadership of palaeontologist and explorer Roy Chapman Andrews (1884–1960). The expedition was originally set up to search for the origins of man in East Asia. The team found many older fossils on site, including the famous nests of dinosaur eggs and well-known species such as *Protoceratops, Oviraptor* and *Velociraptor*, familiar from *Jurassic Park*.

Chapman Andrews was an adventurer at heart. As a student, he dreamed of working for the American Museum of Natural History in New York. When he failed to find a job there straight away, he became a cleaner in the taxidermy department, then an assistant taxidermist. Gradually, he climbed the ladder, until he was assigned to adventurous fieldwork for the museum. 'I wanted to go everywhere. I would have started on a day's notice for the North Pole or the South, to the jungle or the desert. It made not the slightest difference to me,' he wrote in his autobiography, *Under a Lucky Star*. In the end, he made his most famous discoveries in the Gobi Desert. It proved a complex undertaking, requiring him to arrange a convoy of black Ford Model Ts and camels.

In 1930, the recession and political tensions in China and Mongolia forced him to stop his fieldwork. Four years later, his childhood dream came true: he became director of the American Museum of Natural History until his retirement. But retirement did not slow him down. He wrote adventurous books about his fieldwork and dinosaur excavations. 'In the first 15 years of field work I can remember just ten times when I had really narrow escapes from death. Two were from drowning in typhoons, one was when our boat was charged by a wounded whale, once my wife and I were nearly eaten by wild dogs, once we were in great danger from fanatical llama priests, two were close calls when I fell over cliffs, once I was nearly caught by a huge python, and twice I might have been killed by bandits,' he wrote, with some exaggeration. Chapman Andrews is sometimes called the indirect inspiration for Indiana Jones because of his archaeological adventures, a claim that film director George Lucas has always denied.

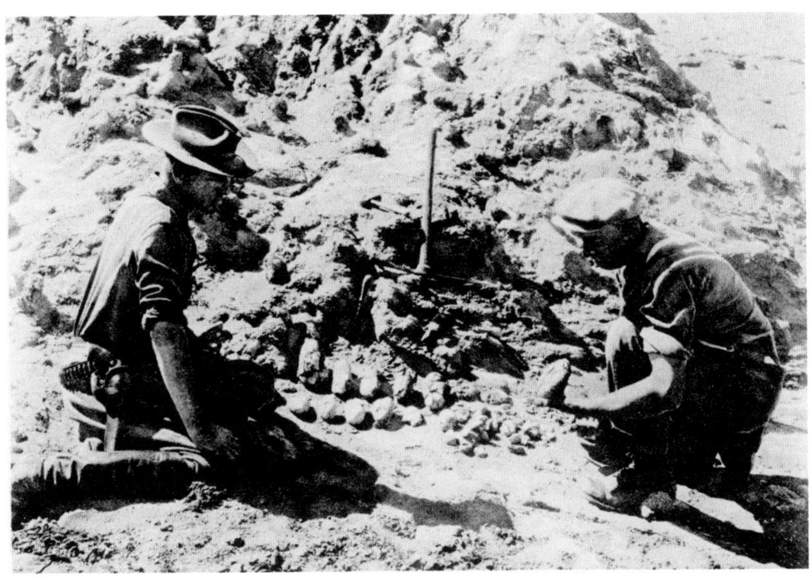

Palaeontologist George Olsen showing his find of fossil dinosaur eggs to expedition leader US explorer and naturalist Roy Chapman Andrews, 1925.

DINOSAUR RENAISSANCE

The 1930s recession and the Second World War forced science to take a back seat. During that murky period, little new research was done on dinosaurs. Additionally, a number of early 20th-century palaeontologists felt that dinosaur science could not be taken seriously. They believed that serious palaeontology involved researching other organisms.

Fortunately, after the Second World War, a period known as the 'dinosaur renaissance' began. It was inspired by John Ostrom (1928–2005) of Yale University, who in 1964 found the first well-preserved remains of a raptor-like animal. He named the animal *Deinonychus antirrhopus* in 1969, after the balancing tail and the terrifying claw on its first toe. Later, he also compared *Deinonychus* to *Archaeopteryx*. Ostrom revived the idea that birds were descended from such raptor-like creatures. His work and that of his students, such as Robert Bakker, rekindled interest in dinosaur science. Research topics such as warmbloodedness, physiology and growth, speed and biomechanics, behaviour, communication and reproduction, evolution and extinction, and many others are the direct result of this palaeontological avalanche. It is still relevant today.

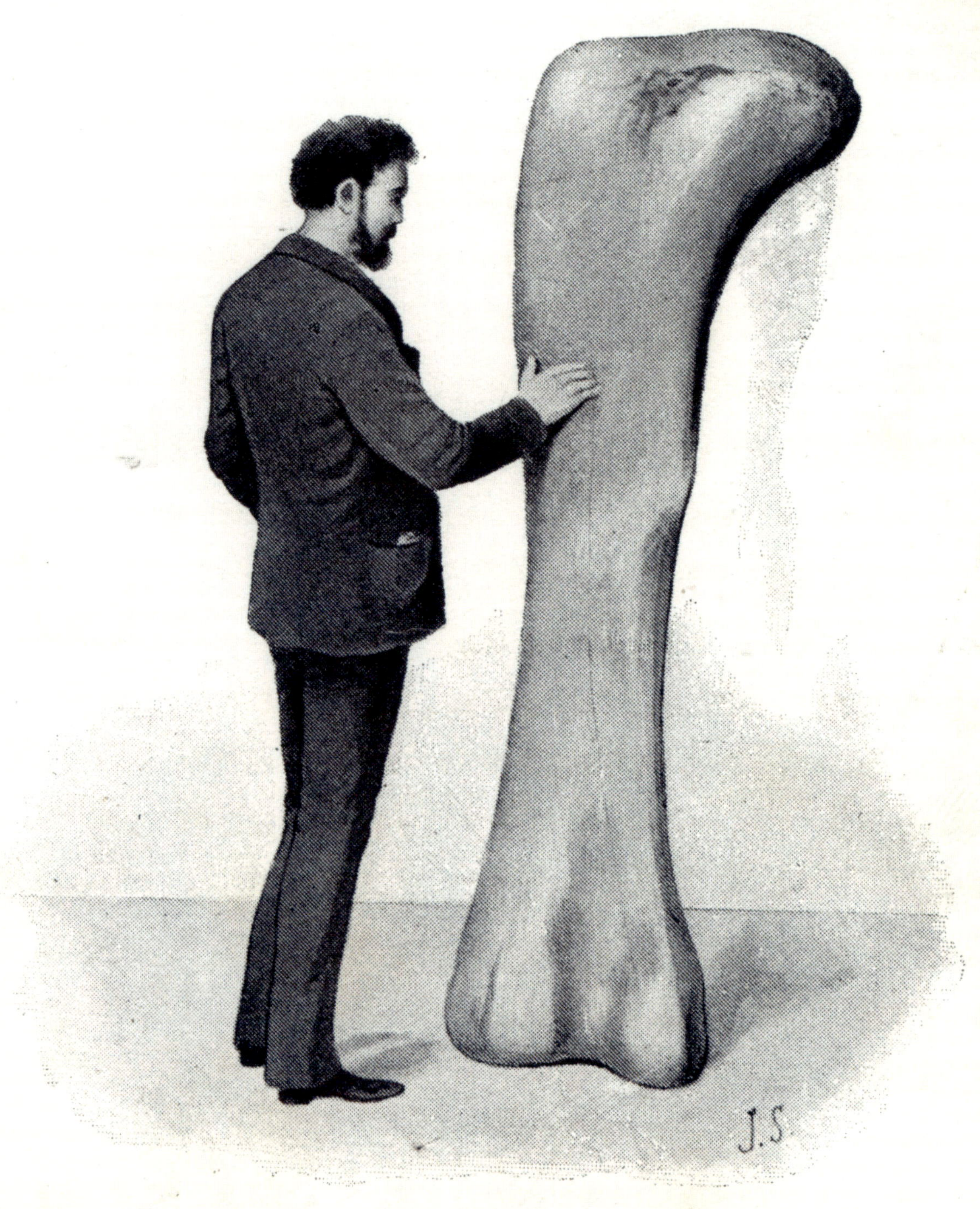

THIGH-BONE OF A HUGE DINOSAUR, ATLANTOSAURUS.

From a cast in the Natural History Museum, London.   Length 6 feet 2 inches.

PLATE XVIII.

CHAPTER 3

FINDING DINOSAURS

FROM DINOSAUR
TO FOSSIL TO DISCOVERY

'Digging up dinosaurs is one of the best things a palaeontologist can do. The field methods haven't changed much for over 150 years – nothing beats a good pair of eyes and some strong shoulders!'

Michael J. Benton in *The Dinosaurs Rediscovered*

A DINOSAUR IS NOT JUST A FOSSIL

The only traces we find of dinosaurs are fossilised remains. Fossils are formed when a plant, animal or other living organism dies, is buried and petrifies. Traces of living creatures are fossils, too. Think of a petrified paw print, an egg, nest or droppings. Statistically, there is little chance of anything becoming fossilised. Remains of this kind can only be preserved for millions of years under extremely specific conditions. Some (parts of) organisms, such as a shell or a skeleton, fossilise more easily than others. A shell in a river will be buried much quicker than the skeleton of a giant dinosaur on land, meaning that the shell has a better chance of being preserved for ever as a fossil. It is exceptionally rare for a dinosaur to turn into a fossil. It only works if the animal dies in ideal circumstances. The corpse must be buried quickly in sediments or sand so that scavengers cannot find the cadaver and eat it. It is also important that the skeleton remains buried long enough. Bones are naturally porous (they are full of small holes) and groundwater can penetrate them. If they are

<

Femur of an *Atlantosaurus*, illustrated in *Extinct Monsters and Creatures of Other Days* from 1894.

left long enough, minerals from that groundwater can crystallise in the pores of the bones. This is how a bone gradually transforms into a stone. In very exceptional cases, the skin or feathers of a dinosaur can also be preserved. If the subsoil contains sufficient oxygen and iron, even cells, blood vessels and proteins from the bones themselves might be retrievable. Researchers dream of isolating dinosaur DNA from that organic material because it would give them a chance to study the genome of dinosaurs – and, who knows, one day even bring a dinosaur back to life (see Chapter 5).

FINDING A DINOSAUR IS PURE FLUKE

To find a dinosaur fossil, you need luck on your side. In the first place, unless you want to dig hundreds of metres deep, a fossil needs to rise to the Earth's surface. This sounds easier than it is. Because the crust of the Earth moves, layers can shift and sometimes push up kilometres of rock. Erosion does the rest. With a bit of luck and patience, the layers containing the fossils are eventually exposed. Palaeontologists want to find out how a fossil came into being. But they also conduct research into the life history of the animal or plant, and, above all, how it fits into our knowledge of evolution. So palaeontology is really a study of death to find out more about life.

HOW DO YOU FIND A DINOSAUR?

It begins with understanding the substrate. This is where geologists, who make maps of the various layers that compose the Earth's surface, come in. To find dinosaurs, you need to hunt in layers of the right age. Dinosaurs existed from about 230 to 66 million years ago and lived almost exclusively on land. The chance of finding a fossil in effluent lava is as good as zero. It's better to look in sedimentary rocks, where the layers consist of rocks deposited by rivers, lakes or seas.

Whether you can find complete dinosaur fossils depends entirely on the geology. In a conglomerate, there isn't much chance of finding a well preserved skeleton. Such rocks form in fast-moving (mountain) rivers or on a beach where there is plenty of tidal action. If an animal dies under those conditions, its body will be broken down, and dispersed by water over a long distance. Fossilisation is impossible. There's a better chance of finding fossils in river deposits with fine rocks like mudstone or shale. These rocks are formed in or along the banks of slower-flowing, meandering rivers with tributaries, channels, distributary channels and adjoining lakes. These are extremely dynamic systems, where the rapid sedimentation of fine material increases the chance of burying more complete dinosaur remains.

THE MORRISON FORMATION: A HOTSPOT FOR DINOSAURS

One of the best known examples of such a fossilised meandering river system is the Morrison Formation in the United States. In its entirety, it covers an area of about one million sq km across the centre of the United States and Canada. The deposits date back to the Late Jurassic, about 155 to 145 million years ago. In the 1870s, the first accidental dinosaur discoveries took place in the Morrison Formation when a new railway line was being built in the area between Wyoming, Utah and Colorado. The finds were not made by trained palaeontologists, who often broke through the bedrock with a blunt axe. They sent the finds, without documenting them properly, by train to museums in Philadelphia or New York, where scientists examined the jumbled fossil remains. In the Morrison Formation, 37 different valid species of dinosaurs have been found so far. In the states of Colorado and Wyoming, fully preserved sauropod dinosaurs such as *Brontosaurus, Diplodocus, Brachiosaurus* and related species, as well as other familiar ones such as *Stegosaurus, Ceratosaurus* and *Allosaurus* have been unearthed. People continue to search the Morrison Formation for dinosaurs. It was also the battleground of what was known as the so-called 'Bone Wars' between Edward Drinker Cope and Othniel Charles Marsh, rival dinosaur hunters who discovered and described dozens of species there (see Chapter 2).

>

A fictional dinosaur herd in the Morrison Formation, the region where railway builders
found the first dinosaur remains in the United States around 1870.

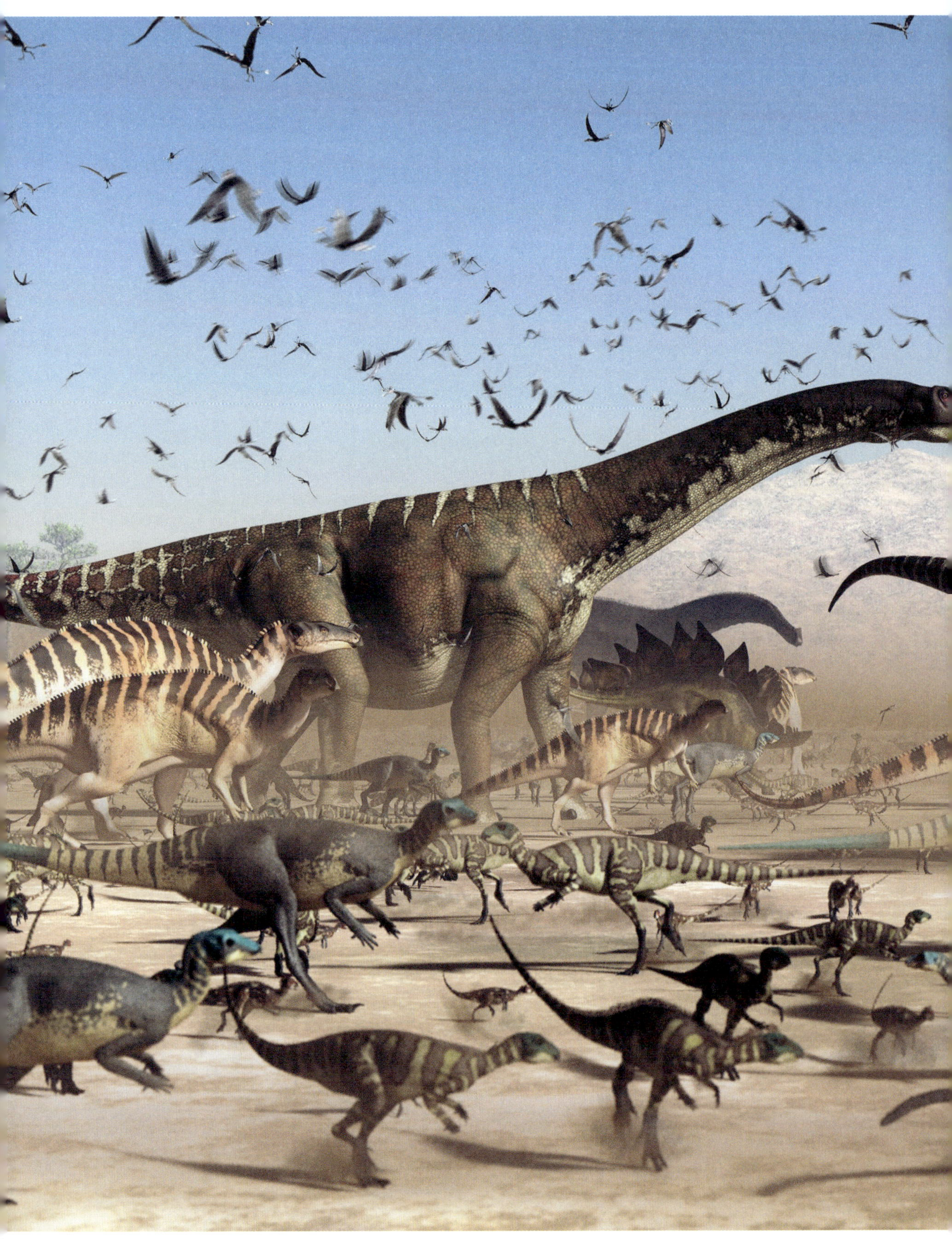

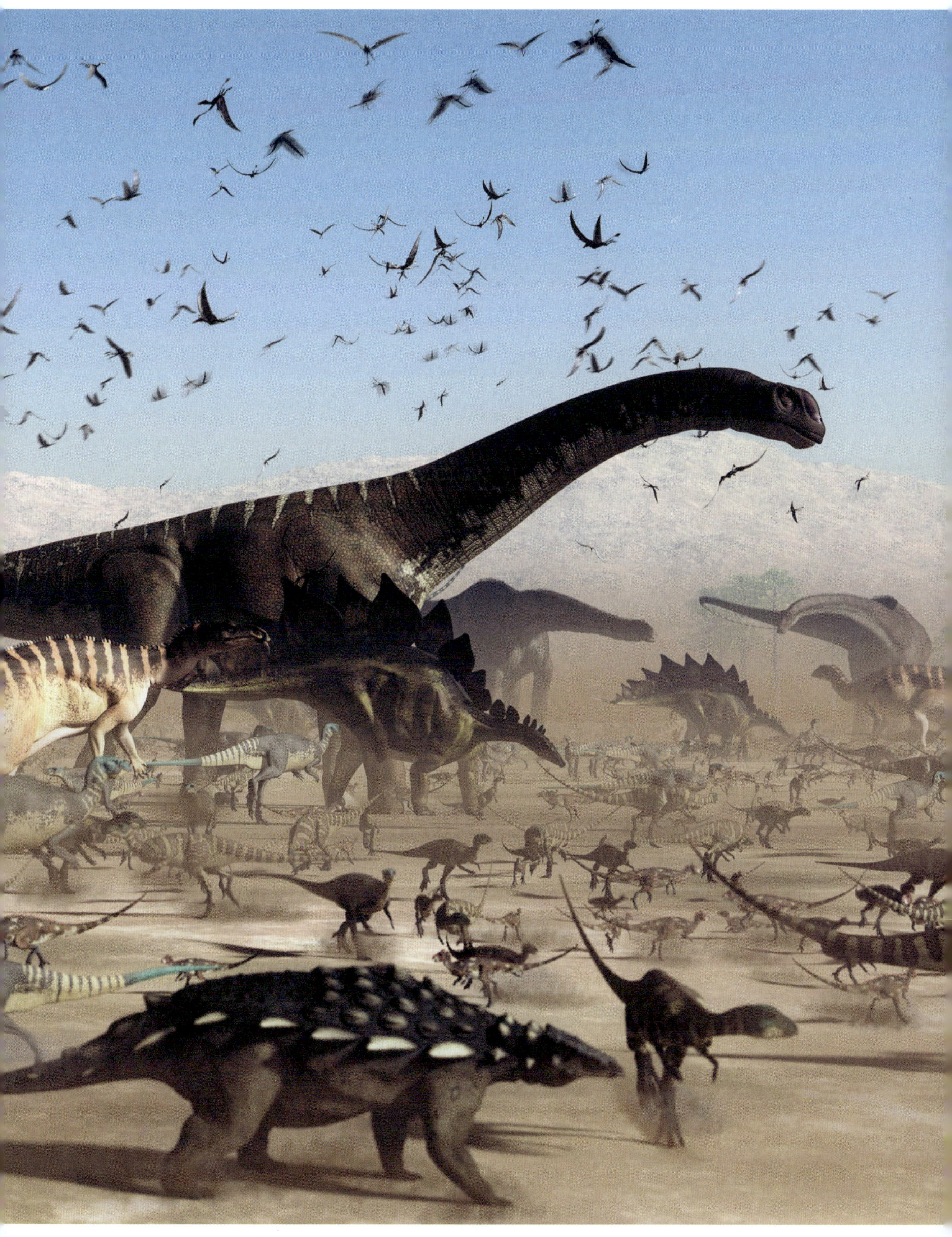

ARE DRAGONS ACTUALLY DINOSAUR FOSSILS?

In his 2001 book on the discovery of dinosaurs, Christopher McGowan talks about 'dragon seekers'. In *The Dinosaurs Rediscovered*, Michael J. Benton writes that the ancient Greeks may have seen skulls of *Protoceratops* in the Gobi Desert that they attributed to dragons. Are myths about dragons based on finds of dinosaur fossils? It would make sense, because all over the planet, stories about dragon-like creatures have been told for centuries, and dinosaurs have been unearthed on every continent. Officially, the first dinosaur fossils were not found until the 17th century. But it's entirely possible that people stumbled across such bones thousands of years ago, without knowing exactly what they were. British palaeontologist Mark Witton explored this geomythology and dragon question in full. On his blog, he published a surprising conclusion: some legends can be vaguely traced back to fossils. But the origin of dragon legends cannot be linked to discovered dinosaur bones.

PARTIAL REMAINS? NO, THANK YOU

In areas that are rich in dinosaur fossils, such as the Morrison Formation or the (Late) Cretaceous deposits of Alberta and Saskatchewan, researchers don't bother to unearth partial remains. A dig easily lasts several weeks and is very expensive. A find must look promising right away, or the crew won't bother to dig it up. This is particularly true of *Hadrosaur* skeletons. The depot of Canada's Royal Tyrrell Museum in Drumheller, Alberta, is packed to the rafters with excavated but unprepared skeletons.

WATCH OUT FOR ILLEGAL HUNTERS

Fossil prospectors are the biggest problem in the dinosaur fossil industry. They operate illegally in countries such as China, Brazil or Mongolia, where fossil exports are banned. Or they don't use scientific methods, so crucial information about the site or geological layer is lost. Prospectors are on the lookout for hunting trophies, and have little interest in the fossil's scientific value. In some cases, they are only interested in the most valuable parts of a skeleton, such as the skull. The rest they leave in the ground undisturbed. Sometimes, finds are tampered with. It's common for a skeleton (or skull) to be assembled from parts from different specimens. Scientists can often tell right away if a find has been meddled with. But unsuspecting collectors, who are offered a skeleton for an attractive price, should be on the alert – if a deal sounds too good to be true, it probably is. The problem is that illegal dinosaur fossils are often traded through middlemen, local people who excavate fossils and sell their finds to traders. They know how to get around rules and regulations in order to sell illegally prospected fossils on the legal market. Laws are complex and vary from country to country. In the United States, for example, you can dig for dinosaurs on private land. In which case, you either own the land and can excavate it yourself, or you have a lease agreement with the land owner which allows you to dig in exchange for a fee.

DORMANT FOSSILS

In 2000, a fossil of *Asteriornis maastrichtensis* was discovered in a quarry near the Belgian town of Eben-Emael. The piece remained in the private collection of a collector for 13 years before he donated it to science. Only then did its scientific importance become clear. In 2020, 20 years after the discovery, Daniel Field published a scientific study on it. The fossil enabled Field to recalibrate the bird family tree. The animal has since been identified, with certainty, as belonging to the Neornithes, the descendants of the last common ancestor of all modern birds. The fossil has features of both Galliformes (chickens and fowl) and Anseriformes (ducks and geese). This means that the split of these groups definitely happened before the mass extinction some 66 million years ago. Modern birds were thus widespread on northern continents early in their evolution, something that could previously only be suspected. The fossil further revealed that birds that survived the extinction did so because of their small size, diet and land-based lifestyle.

>

One Million Years B.C., Tom Liekens, 2018.

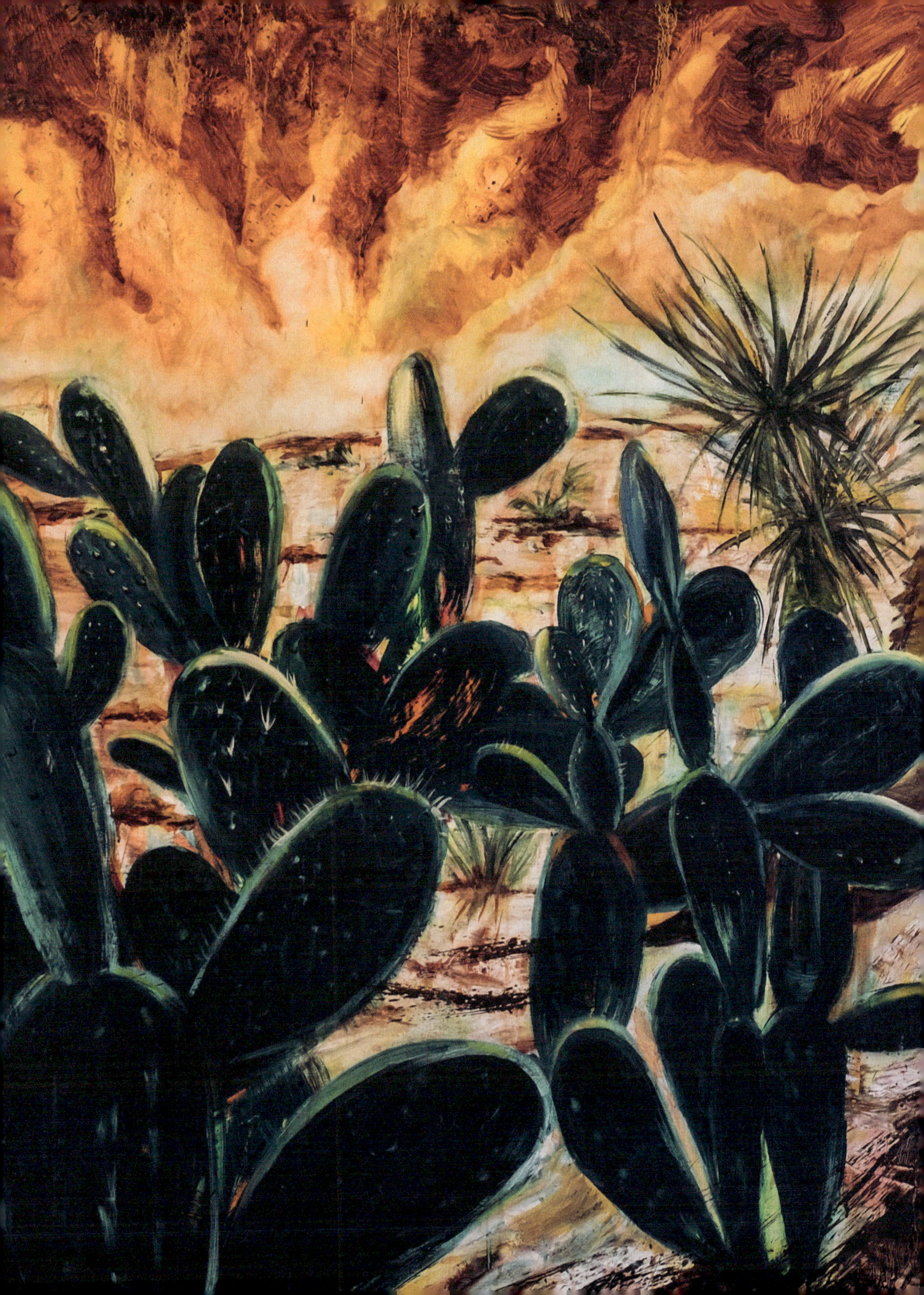

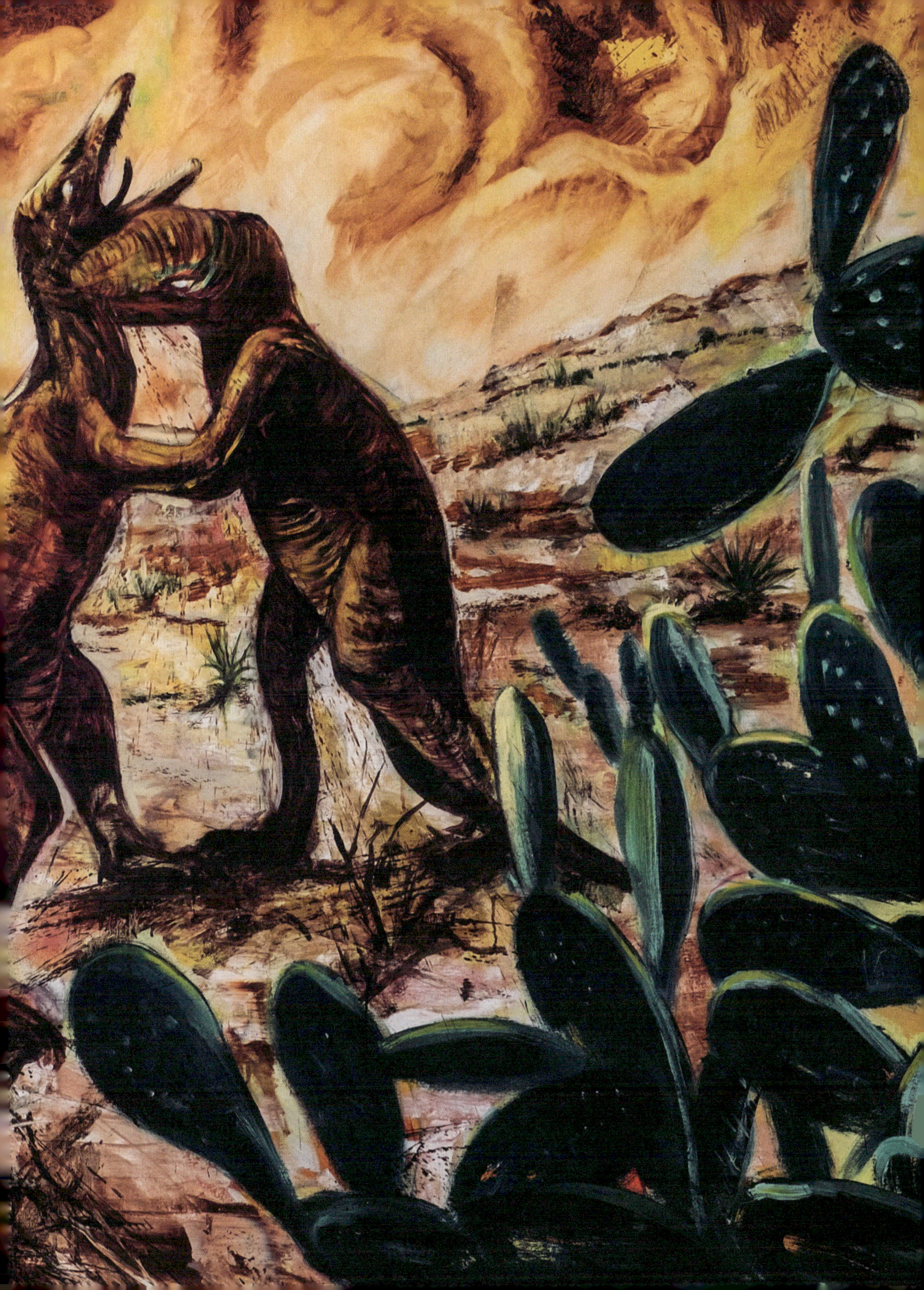

CHECKLIST FOR YOUR DINOSAUR EXPEDITION

1. GO PROSPECTING

Get a geological map and find an area with suitable soil. Makes sure that the right layers are close to the surface, and there aren't too many trees or bushes nearby. First organise a prospecting trip to explore the area. The aim of a prospecting trip is to walk the entire area, scanning for small bone fragments that have weathered from the rock face. 'The secret about prospecting for dinosaurs in the badlands is to look for scraps of bone, and follow them upstream,' writes palaeontologist Michael J. Benton in *The Dinosaurs Rediscovered*.

2. SEARCH FOR TRACKS

With a bit of luck, you'll find a trail of such fragments, which will lead you to a skeleton that is still largely in the same rock face. If you are really lucky, it will be a complete skeleton of a species hitherto unknown to science. Unfortunately, you will only know for sure after it has been fully excavated and prepared in the laboratory so that the anatomy of the bones is clearly visible. You'll often get a good first impression of the type of dinosaur and its completeness at the dig site. Then you can decide whether it is worthwhile to excavate the whole thing.

3. HELICOPTER OR EXCAVATOR

This excavation requires quite a bit of organisation and expense, especially if the excavation site is in a hard-to-reach location. At best, you can work in reasonable comfort by renting a house, cabin or camper van in the area. That way, you don't have to travel too long in the morning to get to the site. If the excavation site is far away, you may even have to hire a helicopter to transport certain pieces as far as the road, where they can be loaded onto a truck. And that, of course, increases the budget. At the site, there is often a lot of rock on top of the layer where the actual skeleton is. Sometimes you even need an earth mover to remove this type of rock. If that's impossible due to accessibility or budget constraints, you'll have to do it with brute force.

4. INVESTIGATING THE CRIME SCENE

Palaeontologists always excavate as large an area as possible horizontally in order to find as many remains as they can. They take into account the structure of the geological layers, the orientation of the bones and the way they are interconnected. Taking good notes is essential, so that later on in the laboratory the right pieces can be reassembled. The aim of the excavation is not to get all the pieces out of the ground as quickly as possible, but rather to uncover them all first, so that you get a clearer overview of what exactly happened so many millions of years ago. It is actually a kind of palaeontological crime scene investigation. The stratigraphic context, the layers in which the skeleton or part of it is found: during an excavation, everything is accurately measured and described. How thick are the layers, how many different types of rock are there, what kind of flow regime was there, what other fossils do we find? These are all questions that will later help answer the ultimate question of what exactly happened to the remains before they were buried. It's also why you need to work slowly and systematically.

5. BRUSHES AND SUPERGLUE

The layer of earth in which you find the bones must always be kept clean. This means using small and large brushes to sweep off dirt and grit and take it to a dump next to the site. This is very important, because you could miss smaller bone fragments, teeth and claws if they were lying under a thick layer of dirt and debris. Or worse, they could be damaged if you step on them. You also need to take litres of superglue to an excavation site. Dinosaur fossils are very fragile. They have often been crushed by the pressure of the overlying rock and are full of cracks. As soon as a bone reaches the surface, you need to fill in the cracks with superglue or another type of glue. Superglue is cheap and available in different viscosities (viscous or very liquid). Glue that flows well can penetrate into the smallest cracks, and viscous glue is excellent for reassembling bone fragments with large fissures.

6. BONES IN PLASTER

Once the bones have been sufficiently and reinforced with glue, remove the rock around them so that the bones form small plateaus. You can then remove smaller, stronger bones from the rock, wrap them in kitchen or toilet paper and pack them with aluminium foil. Larger pieces, such as skulls or the long bones of large dinosaurs, are much too heavy and need extra protection by being wrapped in lengths of jute dipped in plaster of Paris. This technique has been used for almost 150 years to pack fossils at digs and transport them to museums and laboratories. If you are making plaster jackets for larger fragments, you can use strips of wood or metal to add additional support, and embed them into the layers of jute dipped in plaster. The advantage of including wood or metal supports is that they serve as good attachment points for a crane and come in handy when lifting finds weighing several tonnes onto a lorry.

7. PREPARATION AND ASSEMBLY

When the packaged and shipped fossils arrive at the laboratory, the final phase, preparation and assembly, begins. This can take several years. Here the packed fossils are unpacked and extracted from the ground. This process won't take long if the ground is soft and clayey. But it's usually extremely hard, and you need pneumatic hammers to break off the sandstone flake by flake until the bones are completely freed. During this process, the fossils' anatomical traits will emerge, although you won't get a complete picture until all the bones have been extracted. At this stage, the bones might be treated with glue or other preservatives. The numerous cracks in the bones create a great many weak points, which need to be addressed again in the laboratory.

8. DRILLING IN THE THIGHBONE

Once all the bones have been treated, you don't have to worry about touching, turning and moving them as you study them from all sides, measure them, photograph them and, finally, assemble them. Most recent dinosaur studies include drilling a tissue sample from the femur to discover the age of the animal when it died. An important condition for publishing new species is that the specimen is an adult. Exceptions are sometimes made if the skeleton – unusually – is complete. For example, *Sciurumimus*, a carnivorous dinosaur from the Late Jurassic of Germany had an estimated adult length of at least 5m. The holotype skeleton of *Sciurumimus*, however, is no bigger than a cat but 99 per cent complete. Tissue samples can also be useful afterwards for geochemical research into, for example, the state of preservation and the fossilisation process. Ideally, this tissue drilling should be done before the skeleton is assembled.

9. THE DINOSAUR MATRIX

The information about the skeleton's anatomy is described in detail and compared with other species and specimens that may be related to the new fossil. To work objectively, palaeontologists use 'phylogenetic cladistics', a method in which you use a matrix of anatomical characteristics – a kind of checklist – that you fill out. This is the clearest way to compare species. A computer program analyses this matrix to find the most logical relationships between the various species. After that analysis, you can place the new fossil on the family tree. This process identifies the fossil's evolutionary relationship with other species, and, together with the anatomical description of the skeleton, this data forms the backbone of the study, which is eventually submitted to a scientific journal.

10. SCIENTIFIC PUBLICATION

In order to get published, it helps to have information on where the animal was found, its age, how it was preserved, and its age at death. Scientific journals, when they receive a manuscript of a study, will send copies to other scientists who are asked to check the accuracy and veracity of the analyses. They can either reject the manuscript, request a rewrite, or agree to a reworked or unrevised version. Then the article is published, nowadays mostly online, and in the case of some journals, also in print afterwards. The date the article first appears is the official date of the first publication of the new fossil. The entire process from excavation to publication can sometimes take years. As a scientist, you need a lot of patience.

A FORGOTTEN TREASURE
IN THE BACK OF A DRAWER

It is not only fossils from private collections that turn up as 'late' discoveries. The same thing happens in the collections of well-known scientific institutions – especially if they have large collections, such as the British Museum in London. Sometimes, scientists find and dig up far more fossils than they are able to research and describe during their careers. In some cases, fossils may also have been incorrectly described and lie forgotten for years – until a researcher, out of curiosity, opens an old drawer and realises that a fossil deserves a pair of fresh eyes. Such was the case with *Pendraig milnerae*, a small theropod from the Late Triassic Period found in Wales. The skeleton was excavated in 1952 and identified as belonging to an already known species. Angela Milner and Susannah Maidment rediscovered the fossils – which had been lost – in a drawer of fossil crocodile bones in 2020. The following year, Stephan Spiekman and his colleagues officially named the find.

WHAT MAKES A FOSSIL VALUABLE?

People pay a lot of money for fossils, especially now that dinosaur remains have become so collectible. However, fossils have no intrinsic monetary value. The cost price is mainly determined by the materials used and the hours worked to remove a skeleton from the ground, prepare it and assemble it. The financial value of a skeleton is therefore completely separate from its scientific value. A fossil holds scientific value if it is a specimen that was previously unknown or if the find provided fresh information. The discovery of a new species holds great scientific importance because it deepens our knowledge of the biodiversity of the past. A find is also important if it provides new insights from an evolutionary point of view. For example, if you discover that a known species also occurred in a different time period – and is therefore much younger or much older than previously thought – that counts as vital scientific information. It tells us more about the evolution and survival of biological species over time. This is precisely why all the information collected during an excavation is also crucial. Fossils fill gaps in our knowledge of life on Earth and the evolution of the biosphere.

>

The montage of 'Big John', the *Triceratops* that was discovered only in 2014
and sold in 2021 for €6.6 million.

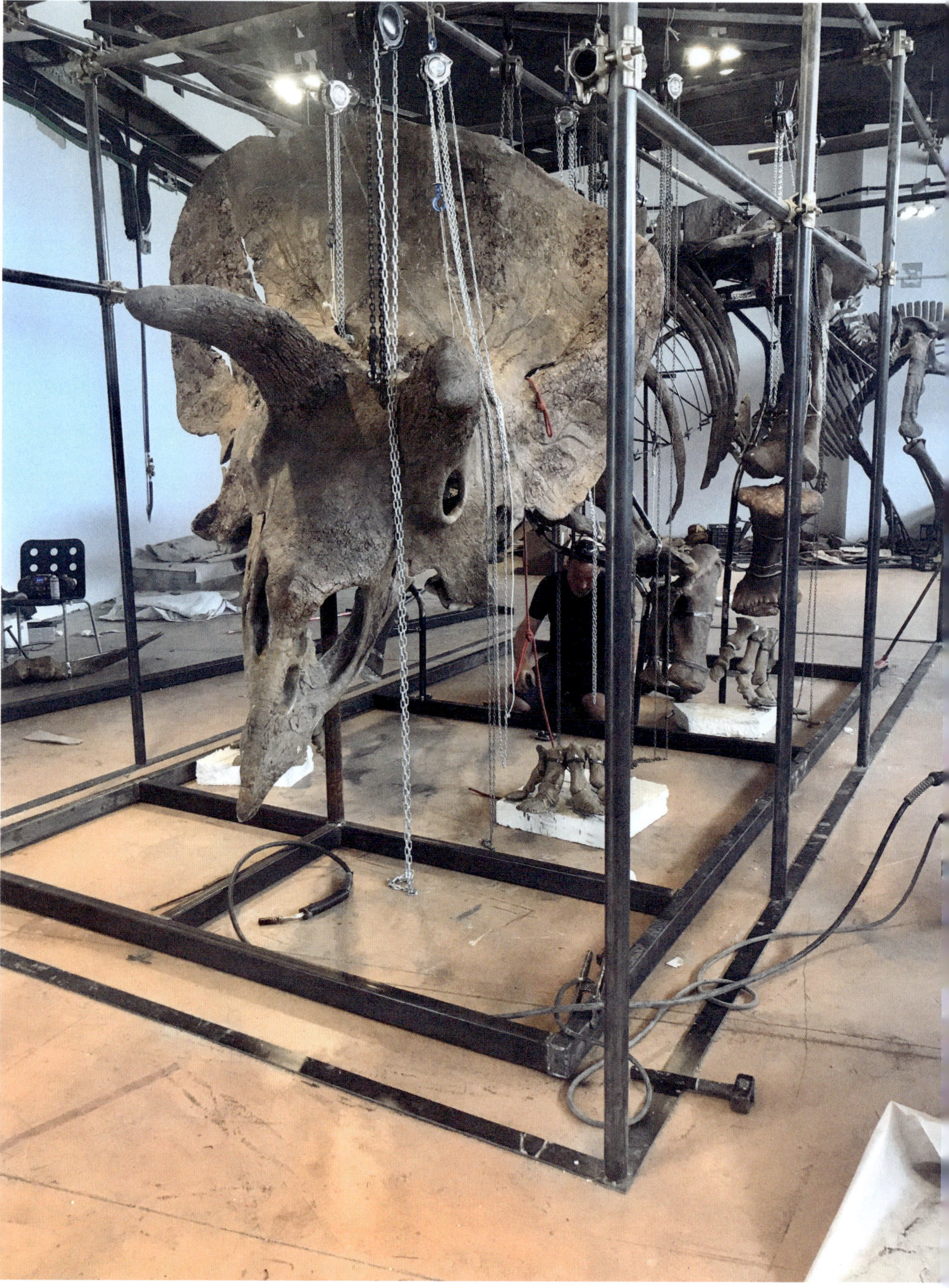

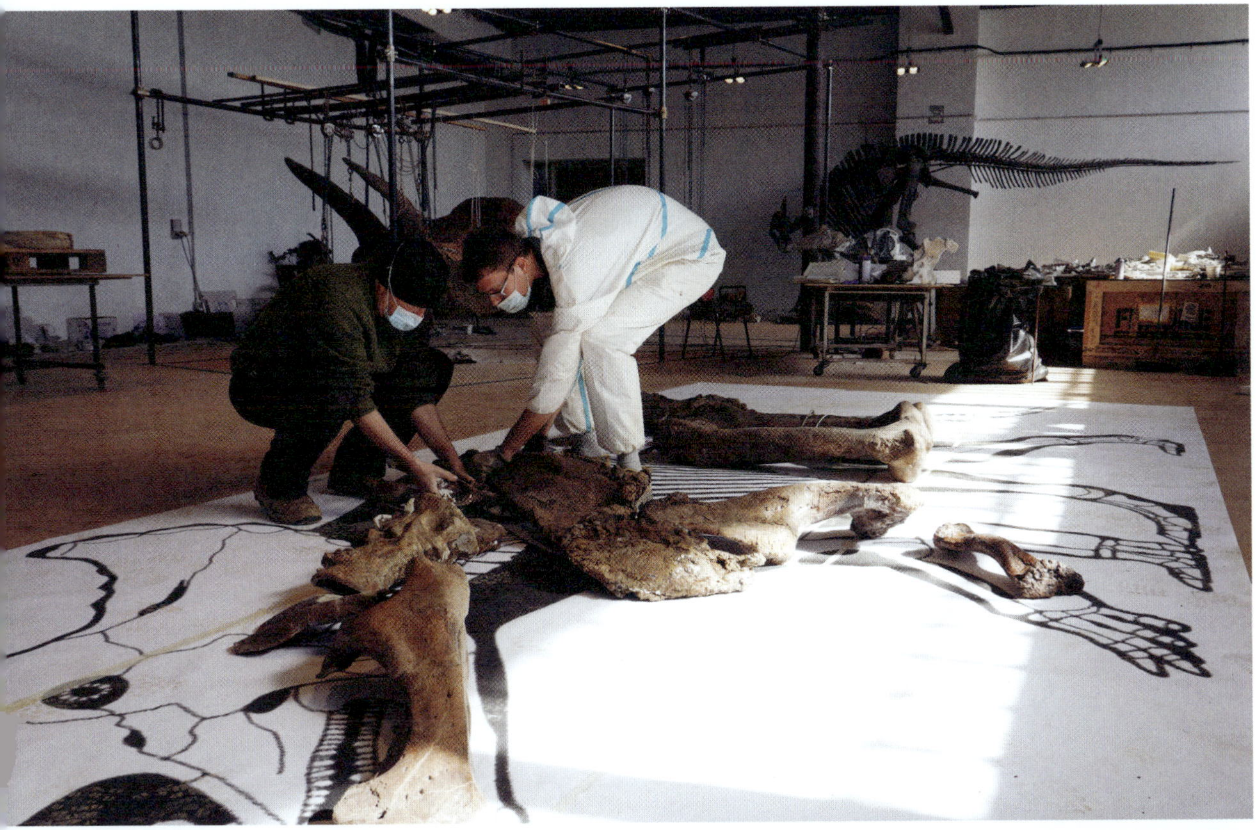

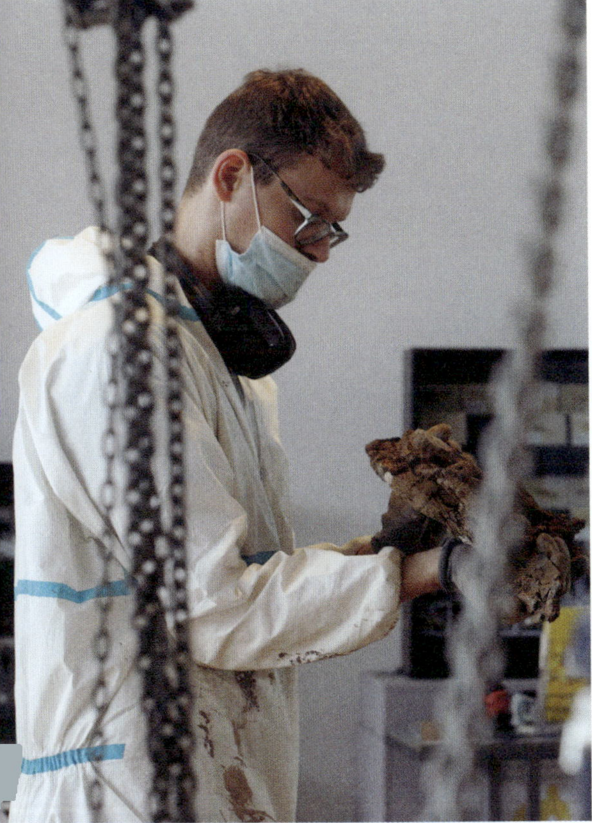
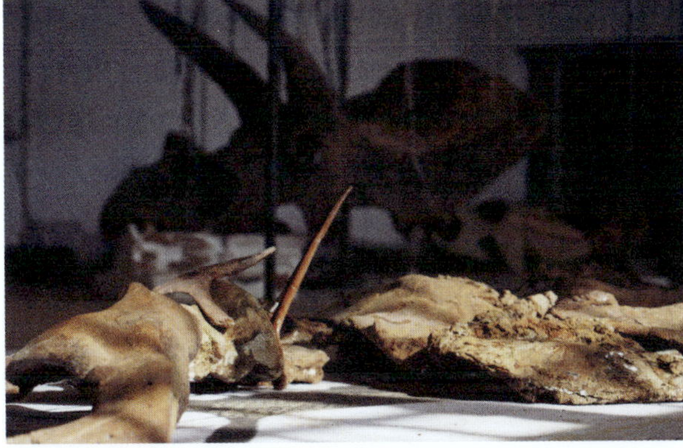
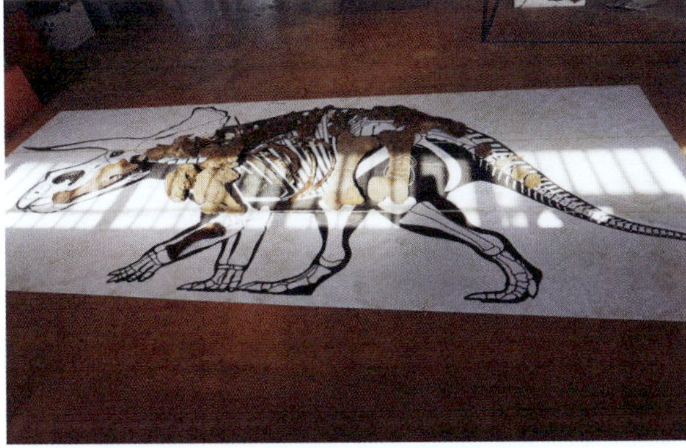

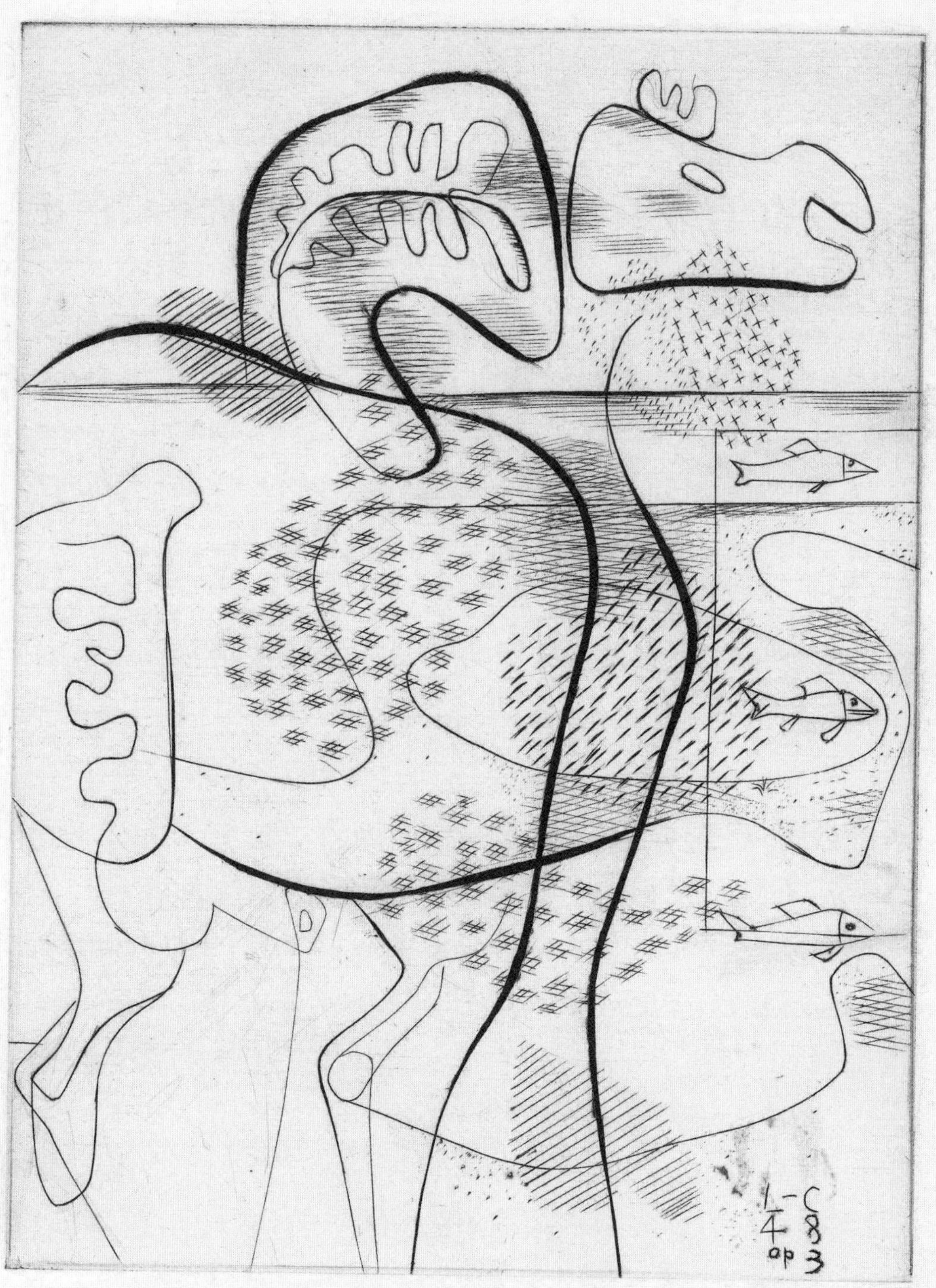

CHAPTER 4

DINOSAURS & ART

PALEOART AND ARTISTIC VISIONS OF PREHISTORY

*'To visualise dinosaurs, you need artists anyway.
There are numerous artists who have drawn dinosaurs,
commissioned by scientific publications or
natural history museums. Without a drawing or
sculpture, these animals will never come to life.
And yet dinosaurs are virtually absent
from the official art circuit.'*

Tom Liekens, artist, 2021

'Mr Martin was deeply interested in the remains of the Iguanodon, etc. I wish I could induce him to portray the country of the Iguanodon. No other pencil but his should attempt such a subject,' observed Gideon Mantell in his diary in 1834. In that very year, the British scientist had once more discovered the fossil remains of an *Iguanodon*, the dinosaur of which he had already found a gigantic tooth in 1822. When the famous artist John Martin visited Mantell in Brighton in 1834, he was shown the fossils. Three years later, Martin effectively created a spectacular dinosaur-themed work of art for Mantell: *The Country of the Iguanodon*. The scene is quite impressive: in an exotic rocky landscape with tree ferns and yuccas, we see iguanodons locked in an epic battle with a *Megalosaurus*. In 1838, the work became the title page of Mantell's important book *The Wonders of Geology* with which both Gideon Mantell and John Martin made history.

<

Dinosaure, Opus 3, Le Corbusier, 1948.

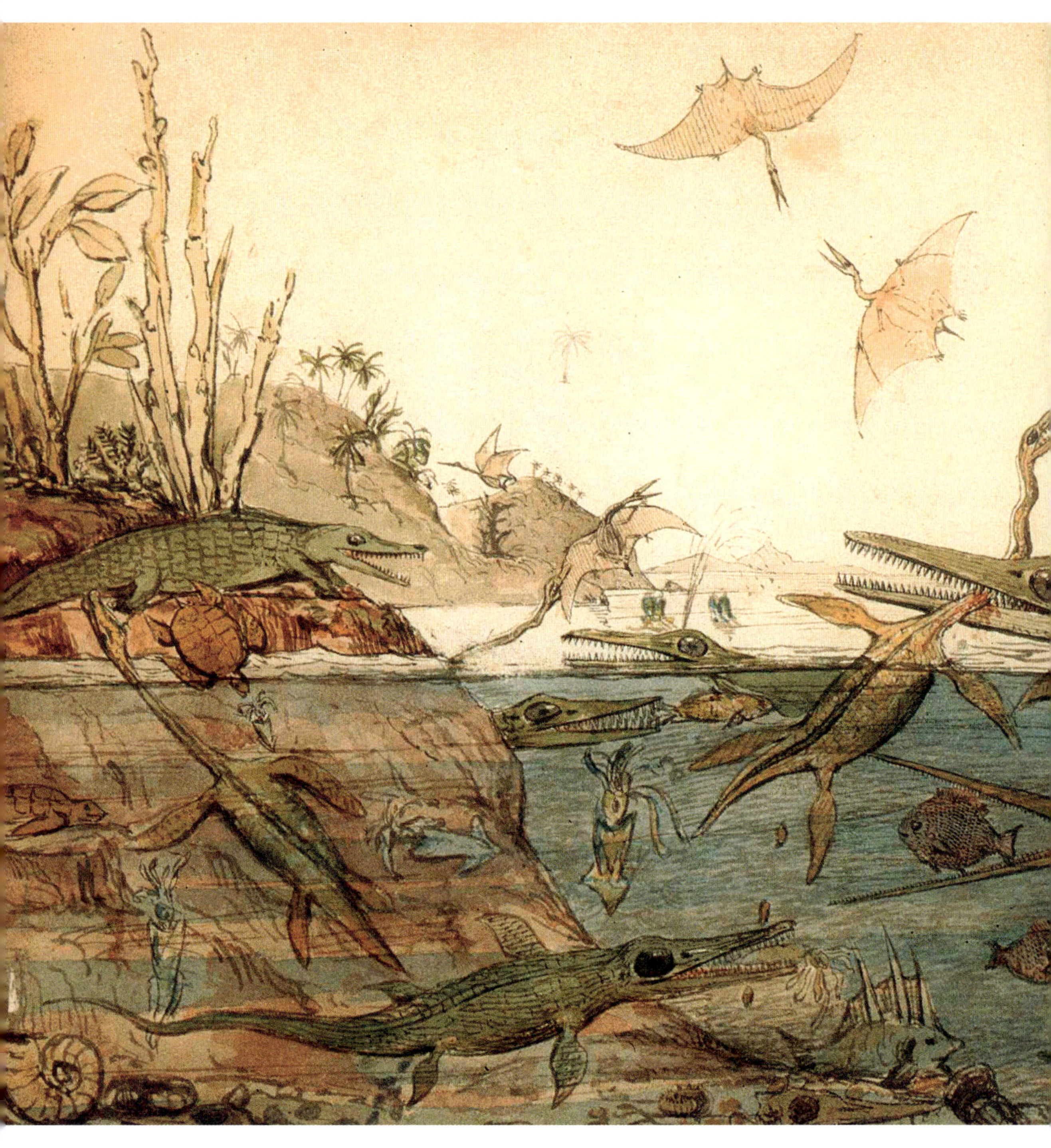

The print *Duria Antiquior* by Henry De la Beche (1830) is the earliest example of paleoart.
The fantasy picture brings together dinosaur species, fossils of which
had been found on the English south coast.

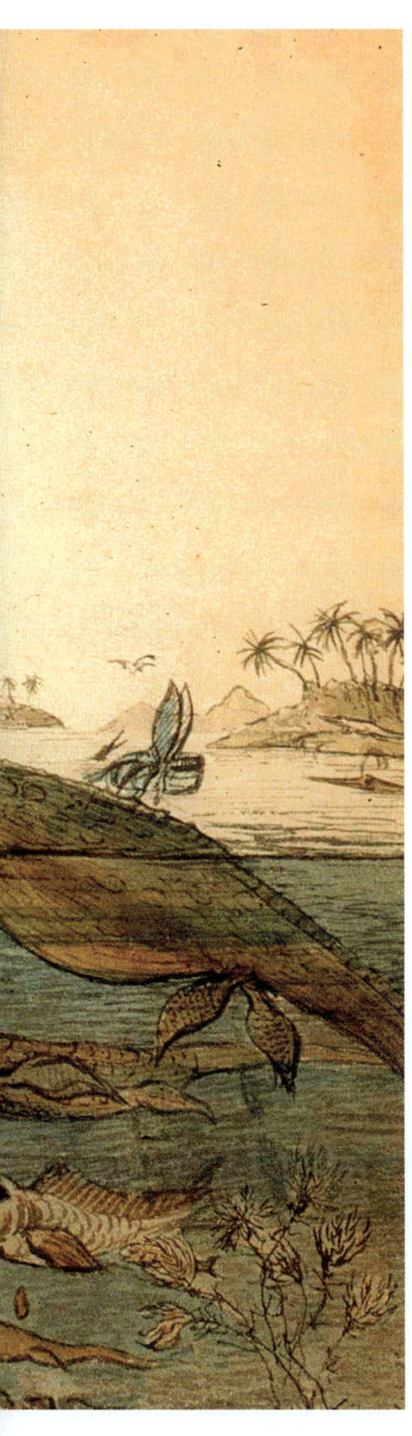

PALEOART AS CROWDFUNDING

With *Duria Antiquior, a More Ancient Dorset*, Henry Thomas De la Beche (1796–1855) laid the foundation for paleoart in 1830. His watercolour, measuring 31 by 22cm, was the unofficial launch of a completely new genre for artists who wanted to depict prehistoric life on the basis of fossil finds. The English geologist's fantasy picture shows the English south coast populated by plesiosaurs, ichthyosaurs and pterosaurs. One plesiosaur even seems to defecate: a reference to the burgeoning study of coprolites, fossilised dinosaur droppings. The scene with the 34 animals is pure fantasy, although it is based on scientific insights. De la Beche based the external features of his dinosaurs on the skeletons that Mary Anning (1799–1847) had found in southern England (see Chapter 2). She was a palaeontologist and fossil collector with whom he was good friends. They both lived in Lyme Regis, met in their teens and shared a passion for geology and palaeontology. Anning not only discovered groundbreaking Jurassic fossils, she also pioneered the study of coprolites. She was not well off and lived on the income from her shop, Anning's Fossil Depot, which from 1826 attracted fossil collectors from all over England. Even Frederick Augustus II of Saxony was a regular customer. Not infrequently, male palaeontologists published scientific studies on the basis of fossils they bought from Anning, without ever mentioning her. This frustrated Anning greatly. 'These men of learning have sucked my brains, and made a great deal of publishing works, of which I furnished the contents, while I derived none of the advantages,' she said.

To support Anning financially, Henry Thomas De la Beche sold prints of his *Duria Antiquior* for £2.10 each, which was quite a lot of money in those days. Nevertheless, the prints became very popular, as it was the first time that dinosaurs had been depicted so vividly. In retrospect, the work is considered the first 'paleoart': art made by palaeontologists or artists to depict prehistoric life. Several versions and copies of the print were made. Some professors at the time even used them in their lectures. The Swiss geology professor François-Jules Pictet de la Rive even had *Duria Antiquior* copied for his 1844 book *Elementary Treatise of Palaeontology*. Piquant detail: the drawing was imitated very precisely, but the droppings were censored.

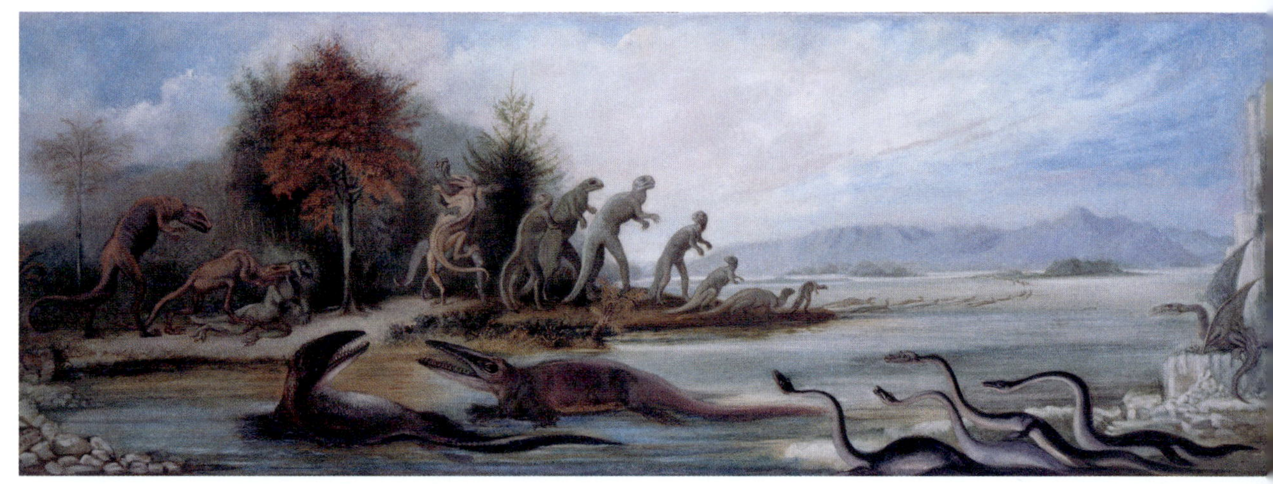

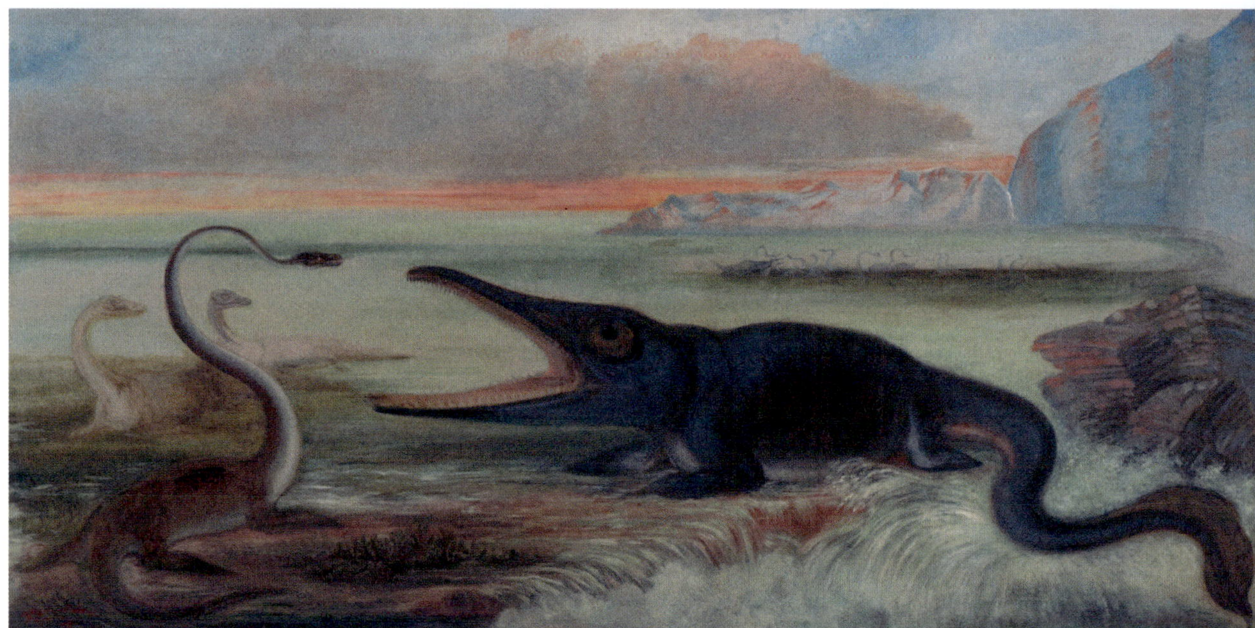

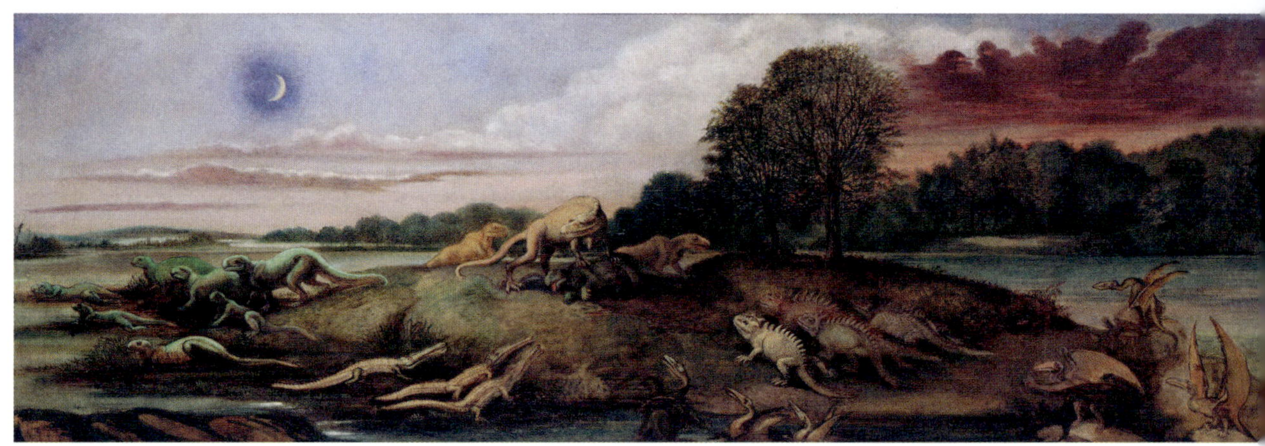

CREATIONIST PALEOART IN PRINCETON

'The greatest scientists are artists as well,' Albert Einstein once said. Benjamin Waterhouse Hawkins (1807–1894) also wore two hats. The way in which he brought prehistoric creatures to life – in sculptures, paintings and as mounted skeletons in natural history museums – had a huge impact on how we still view dinosaurs today. Hawkins inspired our interest in palaeontology. With his prehistoric sculpture garden at the Crystal Palace, he was one of the godfathers of paleoart (see Chapter 2). In 1874, two decades after his pioneering Crystal Palace sculptures (which by that point were already crumbling), he was sought out by Princeton University. Hawkins was commissioned to present American audiences with his vision of prehistoric life – and on a grand scale. He painted a vast panorama, made up of 17 canvases, in what was then still called the College of New Jersey. Fifteen of the paintings survive. The suite of paintings tells the story of the evolution of life on Earth, starting with the dinosaurs. The scenes featuring creatures from the Cretaceous and Jurassic periods are particularly fascinating because they are versions of the scene he proposed to create for the star-crossed Palaeozoic Museum in New York's Central Park. *Cretaceous Life of New Jersey* shows a fight scene between a *Dryptosaurus*, *Hadrosaurus*, *Mosasaurus* and *Elasmosaurus*. *Jurassic Life of Europe* is a tribute to his project for the Crystal Palace and naturally includes a *Megalosaurus* and an *Iguanodon*, which, curiously, stands on four legs. And this despite the fact that, by 1874, scientists had agreed that *Iguanodon* was bipedal (which was challenged only in 1986 by the British palaeontologist David Norman). Hawkins was undoubtedly aware of this, but remained faithful to the vision defined by his mentor Richard Owen in 1854 (the year the Crystal Palace opened).

<

Twenty years after his dinosaur sculptures at the Crystal Palace,
Banjamin Waterhouse Hawkins was commissioned by Princeton University
to paint paintings about the origin of life on Earth.

>

With his 1897 painting *Leaping Laelaps*, Charles Knight was probably
referring to the 'Bone Wars', the palaeontological battle between
Othniel Charles March and Edward Drinker Cope.

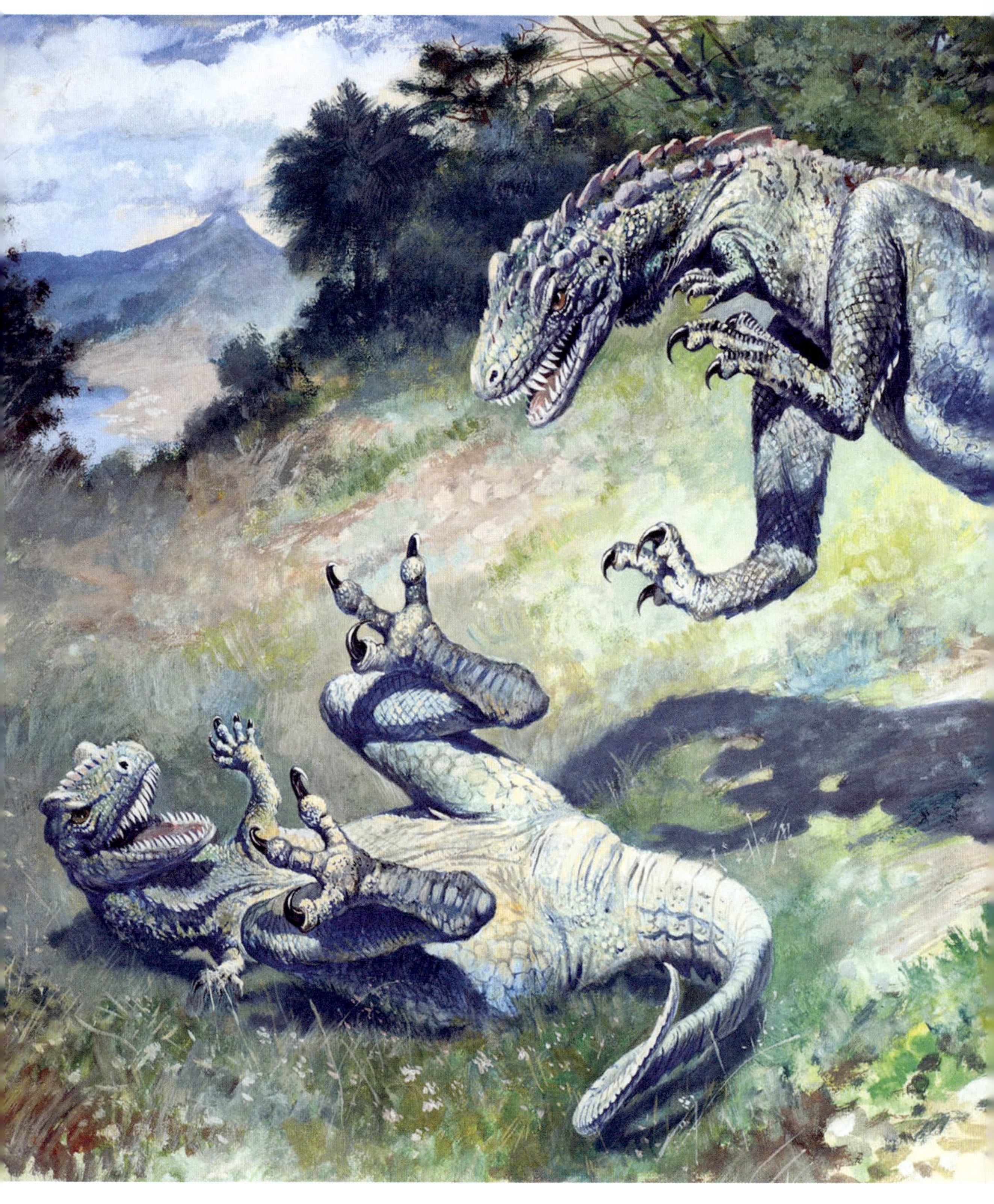

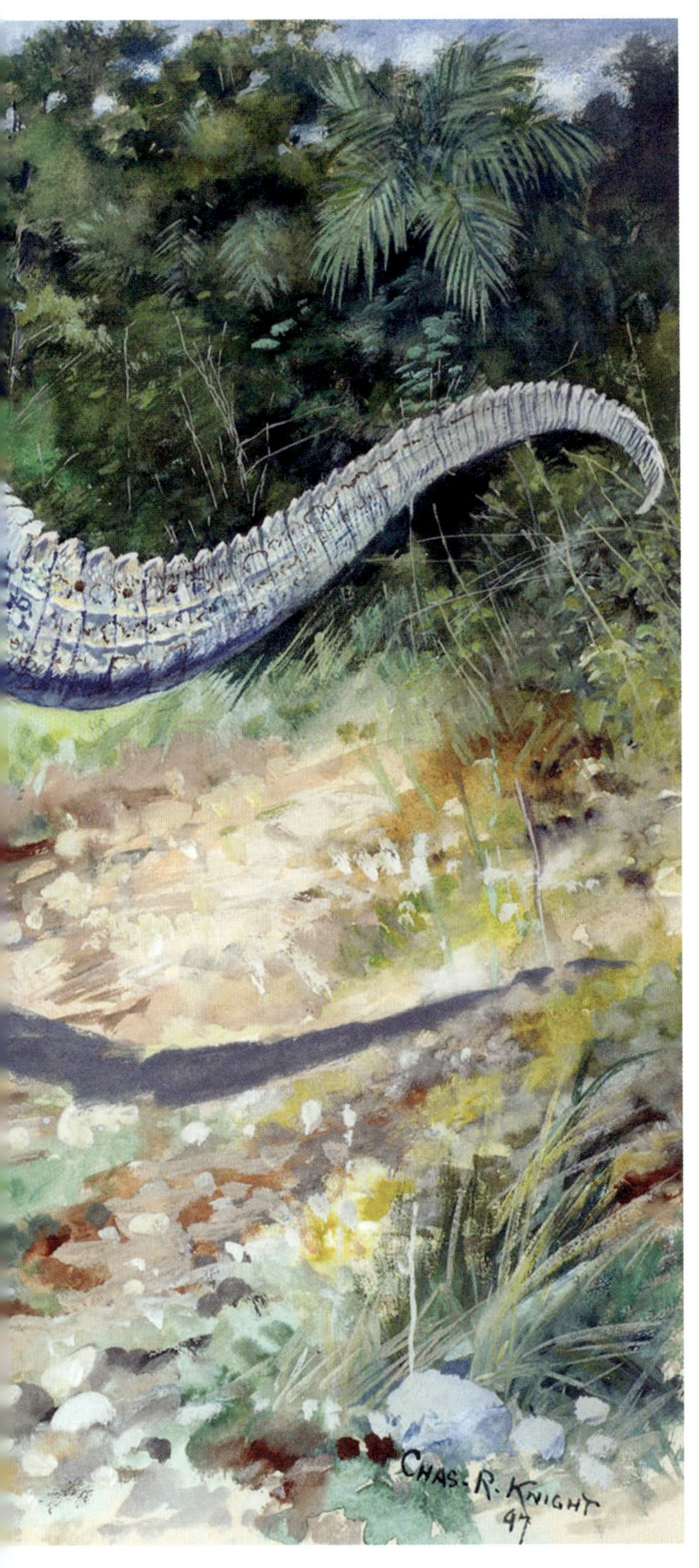

If Benjamin Waterhouse Hawkins (1807–1894) was undoubtedly the most influential-ever paleo-artist, the period of classical paleoart ran from 1890 to 1970, with prominent artists such as Charles Robert Knight (1874–1953), Rudolph Zallinger (1919–1995) and Zdeněk Burian (1905–1981). The most prominent figure was Charles Knight. A much sought-after sculptor and painter, he worked for the American Museum of Natural History, the Natural History Museum in New York and the Field Museum in Chicago, where he painted his masterpiece – a giant suite of 28 pre-historic scenes. In 1897, he first met the notorious dinosaur hunter Edward Cope, a protagonist of the Bone Wars (see Chapter 2). Cope was finan-cially crippled after his disastrous expeditions and mining investments, and fatally ill, but he inspired Knight to produce his first dinosaur paintings, no-table for their lively dynamics and vivid compo-sitions. The best-known early example is *Leaping Laelaps* of 1897, a cinematic scene of two *Drypto-saurus* fighting. Some claim that Knight was actu-ally depicting Marsh and Cope in full fossil-fight mode. The original painting is in the American Museum of Natural History, but the New Jersey State Museum mounted two skeletons in the same pose as Knight's painting. The paleoartist's work is also notable for the almost 'impressionist' nature of his landscapes. Knight did not limit himself to scientific commissions; he also made illustrations for magazines such as *National Geographic* and for the popular dinosaur books by the American pal-aeontologist Frederick Lucas, such as *Animals of the Past* from 1901. An awe-inspiring achievement for someone who was near-sighted from birth. Knight's vision continued to deteriorate until, at 37, he was almost blind. This did not prevent him from taking on huge commissions, for which he hired assistants.

PALEOART AS A (PSEUDO)SCIENTIFIC DISCIPLINE

The Country of the Iguanodon (1837), *Duria Antiquior* (1830) and *Leaping Lae-laps* (1897) are three totally different but world-famous examples of paleoart. Dinosaur art is often found in natural history museums or in universities, as well as scientific publications, encyclopaedias and popular handbooks. 'The lines between entertainment and science, kitsch and scholarship, are often vague,' says Walton Ford in the introduction to Zoë Lescae's book *Paleoart: Visions of the Prehistoric Past*.

Various movements can be distinguished in paleoart. Some artists are professional palaeontologists and strive for the most accurate possible representation of dinosaurs in their biotope. In other words, this branch of paleoart becomes less speculative as scientific insights expand and deepen. Such paleoartists want to incorporate the latest discoveries immediately, for example about the skin colour, plumage or body fat percentage of dinosaurs. Take Gabriel Ugueto, for example. Since a fossil *Plesiosaurus* was found in Mexico with remains of subcutaneous tissue, this artist has been drawing chubbier marine reptiles. This makes sense; as an artist, if you are trying to reconstruct dinosaurs on the basis of fossilised remains, it's tempting to base your illustration on the skeleton and incorrectly draw it much too thin. But if you were to draw a whale or a chicken based on their skeleton, they'd look very scrawny. Palaeontologist Mark Witton, for example, also leaves as little to the imagination as possible in his scientifically based illustrations. 'You need a lot of information to draw a living creature accurately. If my dog were a fossil, I'd need to know a great deal in order to draw it perfectly. I'd need information about the muscles, skin and fur, as well as the skeleton. We don't have much complete information when it comes to dinosaurs. If we still don't know much about a certain species, I prefer to wait until we have more data from scientific papers before drawing it digitally. So why is paleoart art and not scientific illustration? Because I still have to come up with a lot of details myself in terms of composition, lighting and perspective. But I don't want to put too much emphasis on the monstrous side of the creatures in my paleoart. I draw dinosaurs as they would have looked, not as they appear in cartoons.'

Other artists take advantage of the lack of knowledge to let their imaginations run wild and interpret prehistoric monsters freely. In their influential 2012 book *All Yesterdays*, John Conway and many other dinosaur artists advocated for more speculation and imagination in paleoart. They see the art form as a fantasy domain in which to challenge entrenched prejudices about dinosaurs. Vacillating between fantasy, speculation and science, in the 20th century paleoart established a niche for itself in the art world. Important how-to manuals have already been written about paleoart. In 2017, Taschen published a key overview of the history of dinosaur art, and in the last five years, numerous travelling exhibitions have explored this theme, such as those at Teylers Museum in Haarlem and the Paläontologisches Museum in Munich.

WHAT IS THE CONNECTION BETWEEN GOETHE AND DINOSAURS?

Of course, technically speaking, Johann Wolfgang von Goethe (1749–1832) couldn't possibly have heard of dinosaurs: the most important dinosaur fossils discoveries were not made until after his death in 1832. But the poet, philosopher, writer and playwright was unusually passionate about geology, palaeontology and mineralogy. There is even a mineral named after him, goethite. Goethe himself also had an important collection of minerals, rocks and fossils. Between 1780 and 1832, he amassed a collection of over 18,000 pieces. Remarkably, he also had fossils from Weimar, the place where he had lived since 1775. Goethe was well versed in geology and mining and therefore sometimes went on fossil hunting expeditions himself. In January 1819, he wrote to geologist Karl Cäsar von Leonhard: 'We have discovered excellent fossil bones in the region around Weimar. Among them the fragment of a jawbone with teeth similar to those of a *Palaeotherium* [a prehistoric tapir]. As well as the remains of elephants, deer, a horse and other animals at one location.' This may well be true, as the chalky soil around Weimar and Erfurt is known to be a location for finding fragments of prehistoric elephants, rhinoceroses, bison and bears.

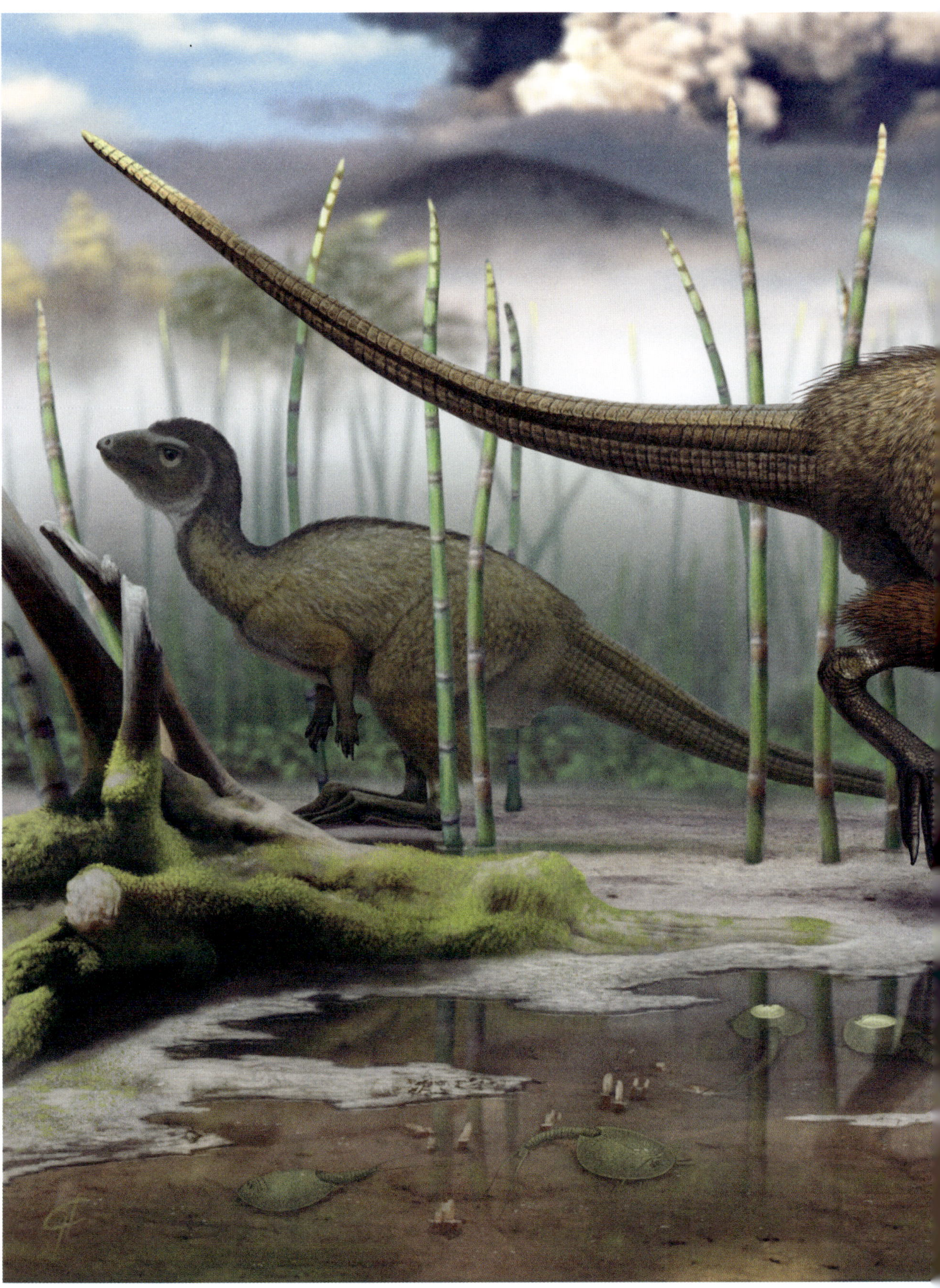

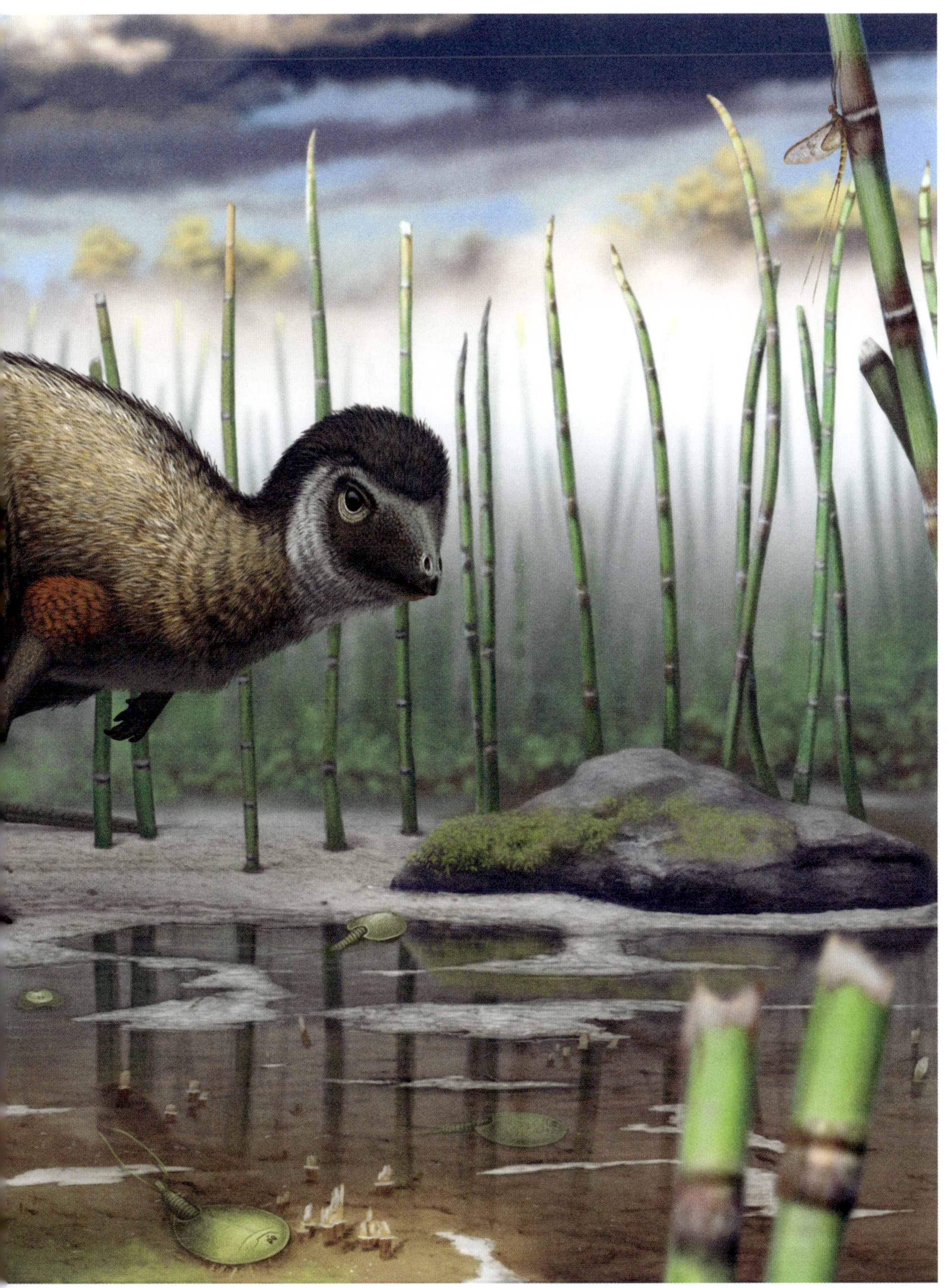

Despite the boom in paleoart, it sometimes seems as if the dinosaur theme has been ignored by so-called 'serious' artists. It's striking how little palaeontology's major breakthroughs, legendary dinosaur finds and dinomania have impacted the visual arts, especially when compared to other 19th-century scientific or historical discoveries such as electric light or photography. The rediscovery of ancient Egypt inspired Egyptomania, which became a phenomenon in the visual and applied arts. In 1853, Japan's isolation came to an end, giving rise to a tsunami of Japonist influences in the Western arts. Excavations in ancient Rome and Greece saw the birth of neoclassicism in painting, architecture and design. And the colonial expeditions in Africa furnished the first African art collections in museums, which inspired countless artists of the time. Pablo Picasso, Man Ray, André Breton and Emil Nolde; all fell under the spell of these 'primitive' art forms, each artist integrating them into their artistic endeavours in their own unique style. Oddly enough, the palaeontological breakthroughs and dinomania of the same period barely made an impression on 'high' art. Despite fantasy stories about human encounters with 'lost' dinosaurs that filled newspapers and magazines. Despite the 1925 stop-motion dinosaur film *The Lost World* that scared countless moviegoers witless. And despite the first dinosaur skeletons in natural history museums that drew tens of thousands of visitors. Artists such as Van Gogh, Picasso and Matisse must have been aware of the breakthroughs in palaeontology, but never referenced these discoveries in their work. Dinosaurs appear sporadically in 20th-century modern art history. One of the best-known portrayals is Le Corbusier's etching *Dinosaure, Opus* 3 of 1948. In Paul Klee's lyrical bestiary, one can – if feeling generous – sometimes make out prehistoric animals. But in modern art, there are certainly very few portrayals of dinosaurs.

<

A *Kulindadromeus zabaikalicus* illustrated by paleoartist Andrey Atuchin.

DINOSAURS IN CADILLACS

Kenny Scharf (1958) was a contemporary of Jean-Michel Basquiat and Keith Haring in 1980s New York, who fortunately outlived his late colleagues. Like Haring and Basquiat, he wanted to liberate art from its elitist confines and did so by sprinkling references to popular culture, from *The Flintstones* to *The Jetsons*, from kitsch toys to monster movies and science fiction throughout his work. In the American cartoon series *The Flintstones*, prehistoric people live in harmony with dinosaurs. Fred Flintstone has a pet dinosaur. As a crane operator, he sometimes rides around on a *Brontosaurus*. And when they are hungry, they eat *Brontosaurus* burgers or ribs.

With his 'dino works', Scharf is also on the cutting edge of prehistory and popular culture. In a vast mural in New York's Bronx in 2016, he incorporated a lazy *Brontosaurus* as well as aliens and other anthropomorphic creatures. Scharf's playful compositions in euphoric colours capture the dynamism of the big city and the visual miscellany of our saturated visual culture. The artist doesn't just paint murals; he also makes sculptures, installations, paintings and even dips a toe in fashion and design. For decades, dinosaurs have been a signature motif in his colourful visual world. In 1985, he made the *DinoPhone*, a Pop art sculpture with a toy phone and plastic dinosaurs. In 2012, he pimped a 1959 Cadillac into his *New and Improved Ultima Suprema Deluxa*, including a dashboard with toy dinosaurs and characters from *The Flintstones*. 'A lot has to do with the future and with the past. I always combine dinosaurs, which I like because they connote fuel and oil and gasoline consumption. I like rockets and jets and science fiction. I like having dinosaurs eating jets in the cars,' says Scharf.

———

>

Dino Phone, Kenny Scharf, 1985.

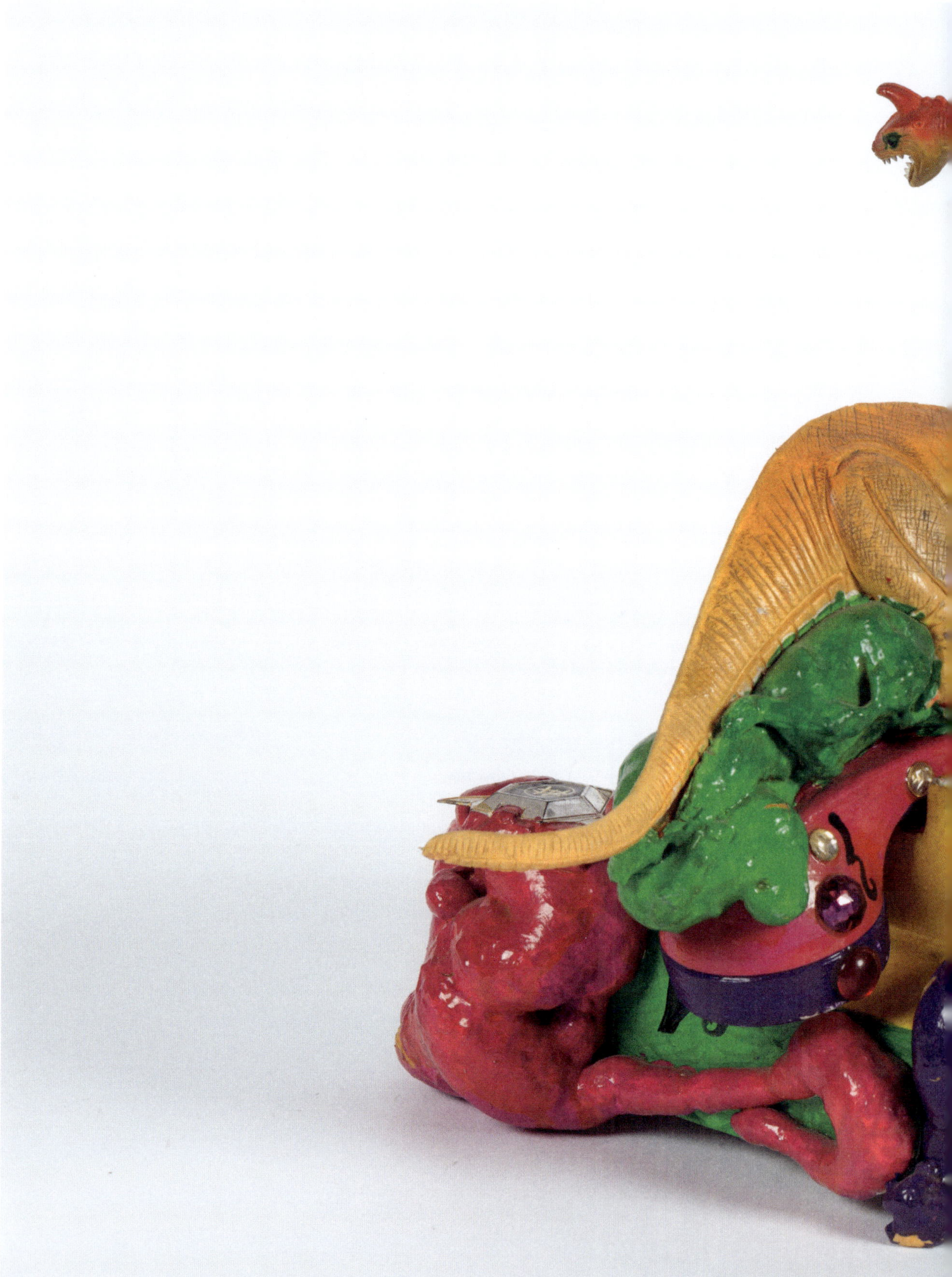

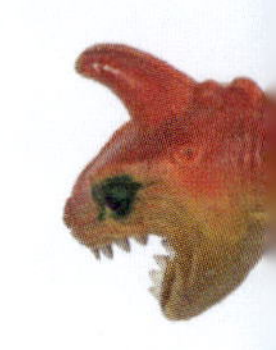

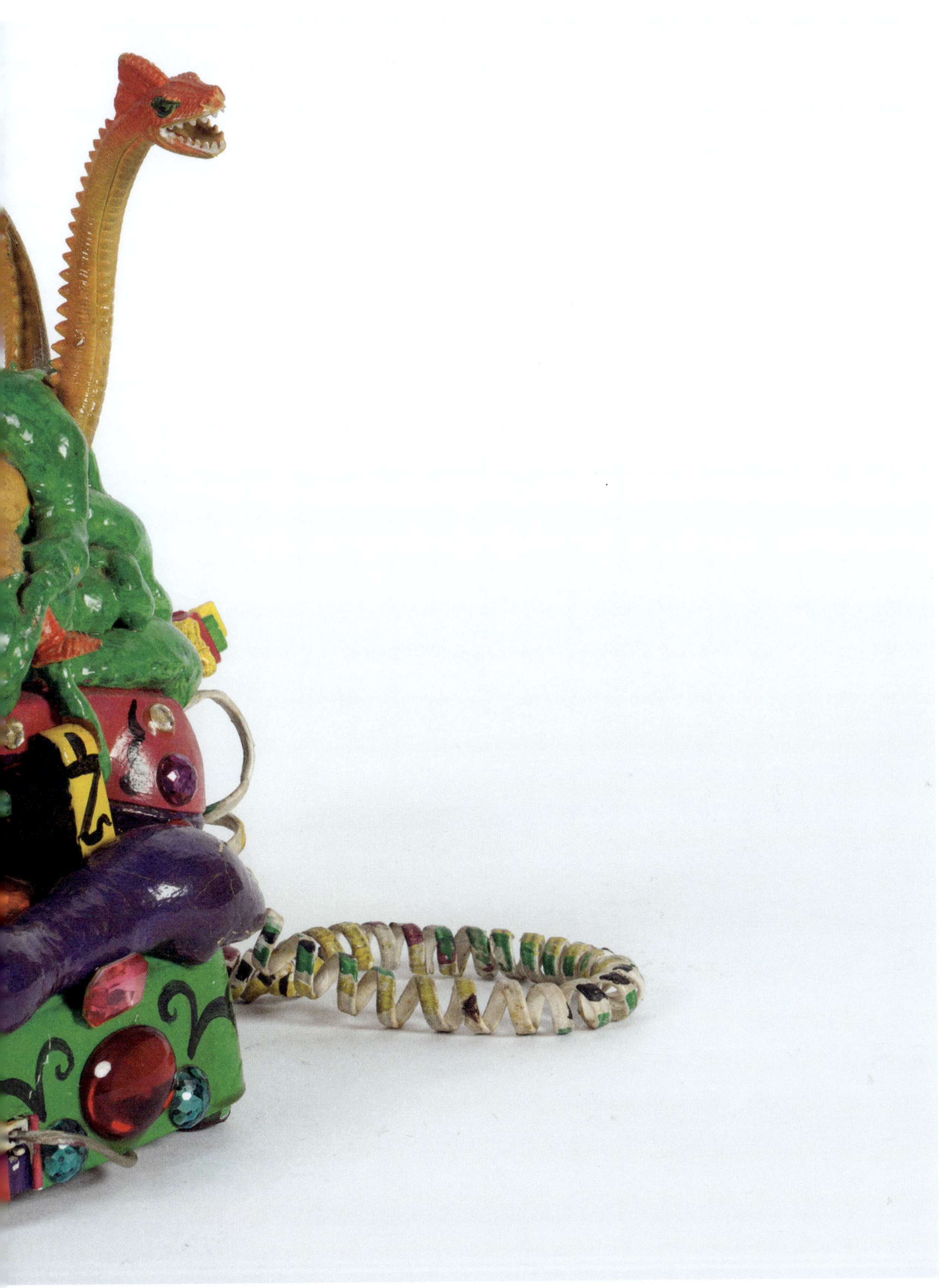

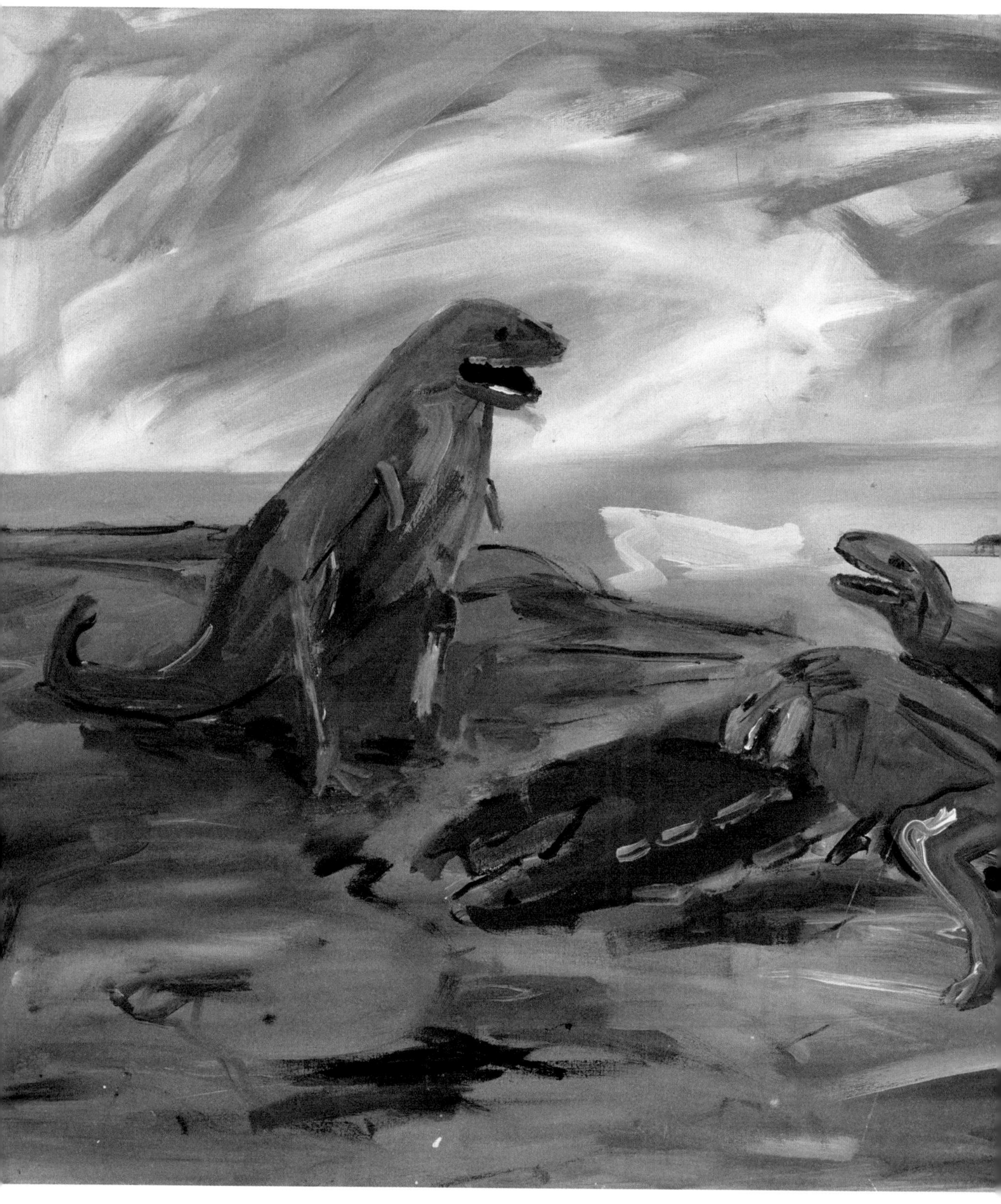

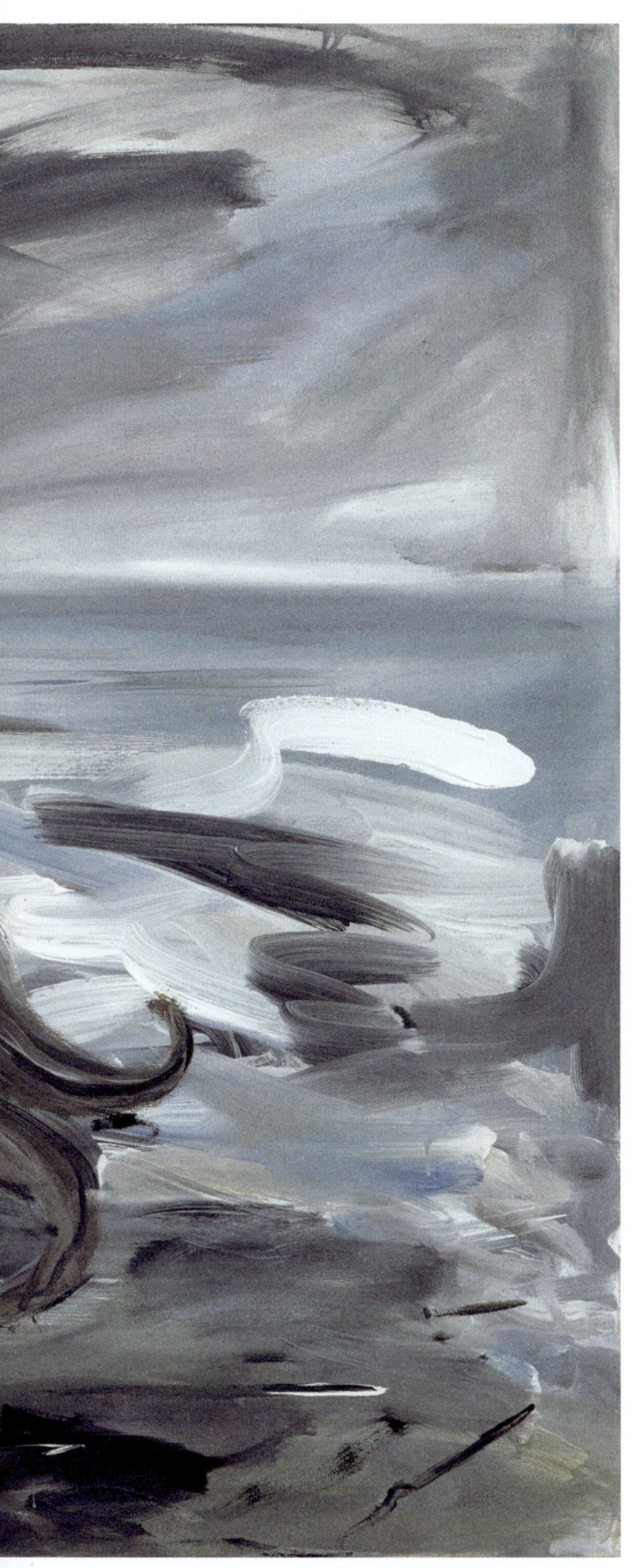

HASTY DINOSAURS

In the early 1980s, the German artist Albert Oehlen (born 1954) made several hastily executed paintings with dinosaurs as the underlying theme. *Barbecue* and other paintings are painted with an intentional casualness, with a grey-toned abstract-expressionist background and a 'banal' figurative image in the foreground. Oehlen's works from that period push the boundaries of abstraction and figuration. In his oeuvre, he embraced 'Bad Painting' – a nihilistic quest in search of the meaning or futility of painting, a medium that has its roots in 'prehistory'. Oehlen's portrayal of dinosaurs is secondary to the artistic questions he implicitly asks: As an artistic discipline, does painting still have a purpose? What should a painter be doing today? Or, what makes a painting good? In these early works, Oehlen deliberately allows a certain awkwardness into his paintings. He doesn't believe that painting needs to be an exalted activity that requires accuracy or talent. A painting can just be bad. By deliberately making sloppy, meaningless and hasty works, he dismantles the medium of painting, stripping it back to the bone.

—

Barbecue, Albert Oehlen, 1981.

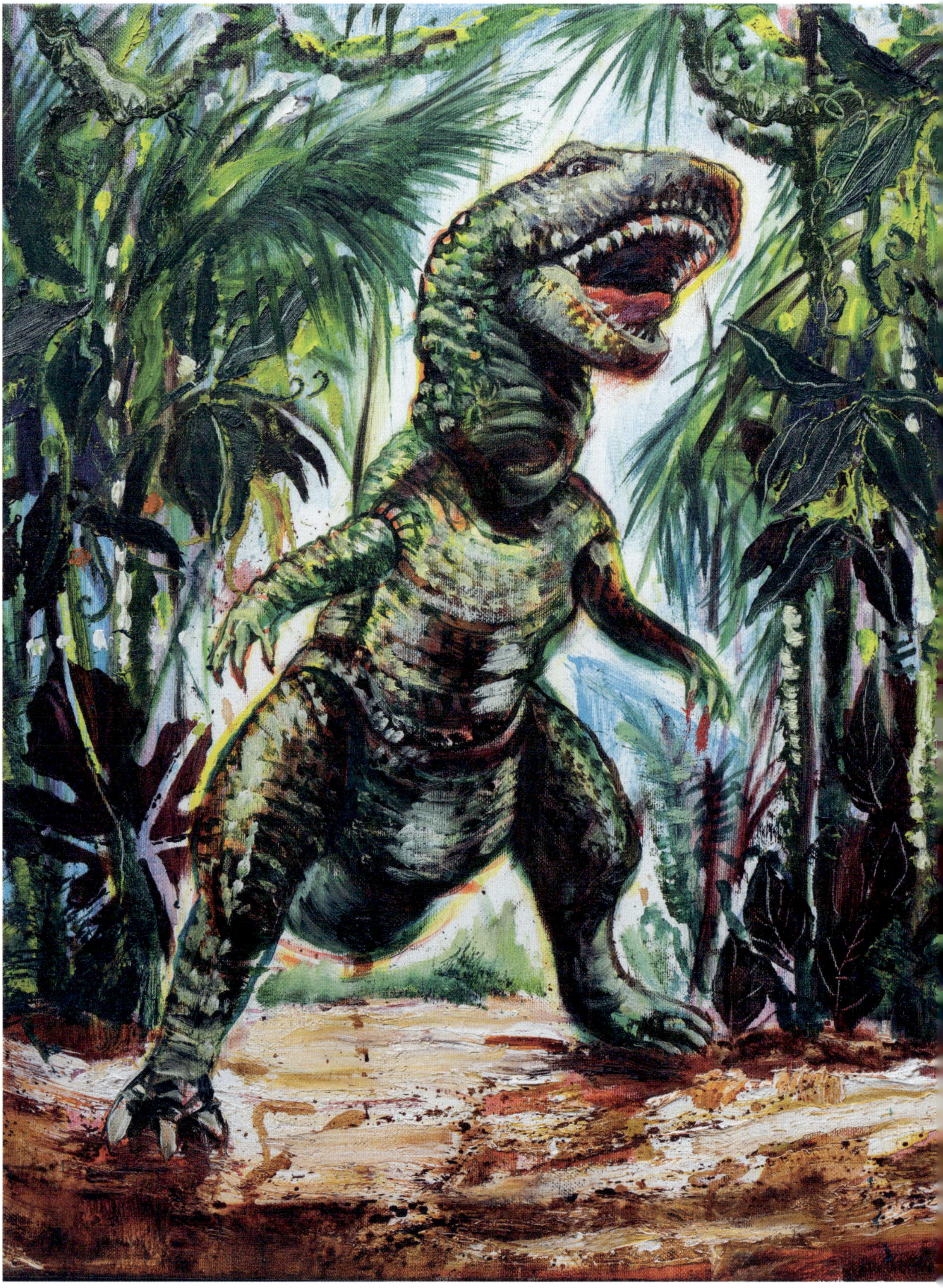

TOM LIEKENS'S WOODEN DINOSAURS

'In his artworks, Tom Liekens plays with the ambiguous prehistory of the supposed coexistence of humans and dinosaurs, and with the evolutions in the depiction of extinct species. It illustrates how volatile our knowledge of prehistory is. And how strongly worlds, imaginary and real, can change,' writes biologist Dirk Draulans about the work of Tom Liekens (1977), who besides being a painter and printmaker is also a passionate collector of fossils, naturalia and stuffed animals. As an artist, he says he is particularly inspired by the representation of nature in science and popular culture. In 2017, he embarked on a series of works portraying dinosaurs and prehistoric animals. Liekens is fascinated by how dinosaurs were represented in the 19th century in museums and in Hawkins's theme park at the Crystal Palace. 'In Victorian times, creationists believed that dinosaurs were monsters that lived on earth before the Flood. The discovery of dinosaur fossils must have led to all sorts of myths and legends about dragons or prehistoric giants before the great palae-ontological breakthroughs,' he says. Liekens also looks at the representation of prehistory in toys or in legendary dinosaur films such as *The Lost World* (1925), *King Kong* (1933), *Godzilla* (1954), *One Billion Years BC* (1966) and *The Land That Time Forgot* (1975). 'All those films are always about the rela-tionship between humans and dinosaurs. While humankind has never wit-nessed this prehistoric fauna, in popular culture, we stubbornly stick to this anachronism,' he observes.

The dinosaurs in Liekens's paintings or works on paper are rather wooden creatures and the landscapes in which they figure are mostly imaginary. Liek-ens stages epic battles between dinosaurs and gorillas, like King Kong. His paintings almost look like film stills, which say much more about how we look at dinosaurs than about how scientists investigate prehistory. 'To visualise di-nosaurs, you need artists anyway. There are plenty of artists who have drawn dinosaurs, commissioned by scientific publications or natural history muse-ums. Without drawing or sculpture, these animals do not come to life. And yet dinosaurs are virtually absent from the official art circuit,' he says. 'When I started a series of paintings and works on paper about dinosaurs in 2017, I initially felt I had to cross a threshold. I had the feeling I was treading a fine line with illustration and kitsch.'

———

<

T.rex, Tom Liekens, 2018.

>

Iguanodons, Tom Liekens, 2018.

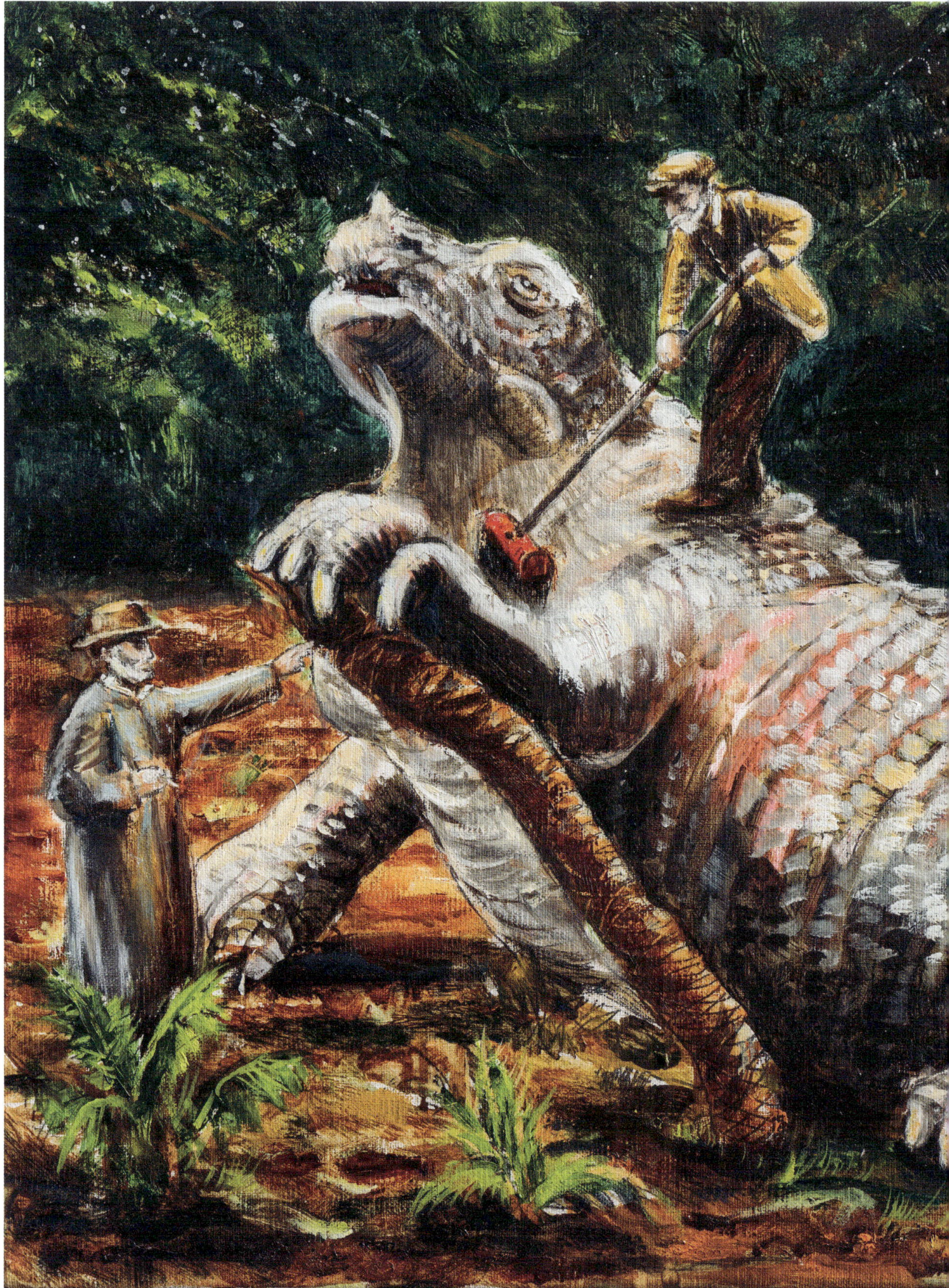

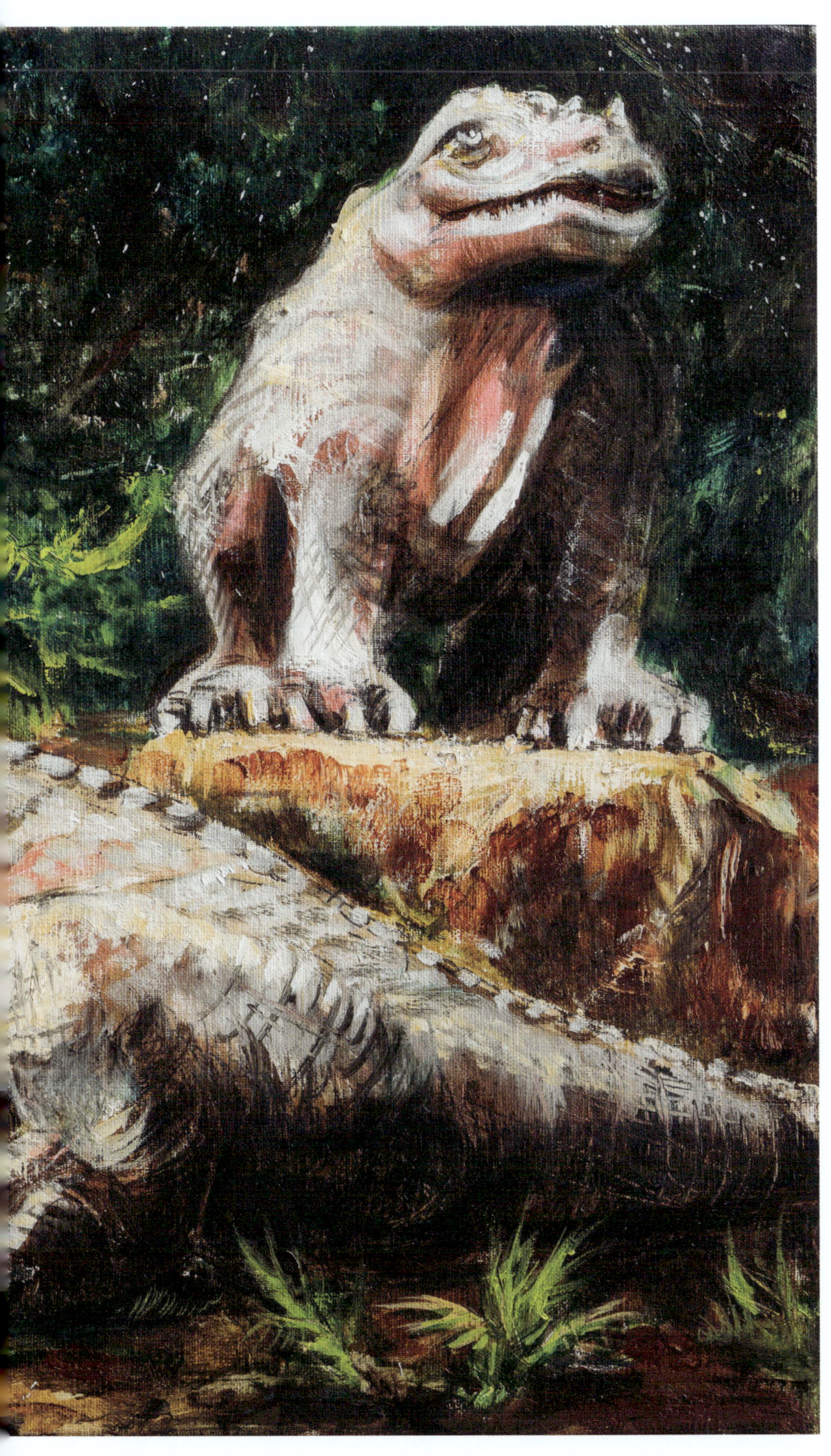

CARTOONISH DINOSAURS

Dinosaurs are also a recurring motif in the work of Jake (b. 1966) and Dinos Chapman (b. 1962), the British brothers known for their shocking sculptures that explore immorality, violence, religion, war and other darker depths of humanity. Their grotesque contemporary works of art do not shy away from controversy or horror. For the Chapman Brothers, dinosaurs are symbols of an uncivilised life of brutality, long before there was any talk of so-called 'human civilisation'.

Explaining Christians to Dinosaurs was the provocative title of their solo show at Kunsthaus Bregenz in 2005. In the etching of the same name, the Chapman Brothers depict two clumsy fighting dinosaurs, with a comet and a volcano in the background. Making a stronger impression is their series *Hell Sixty-Five Million Years BC* (2004–2005) and *Hell 63 Million Years BC* (2009), a series of dinosaurs made from painted cardboard, exactly as a toddler would. For a kid, the prehistoric creatures are either bloodthirsty killers or cute cuddly pets. The Chapman Brothers portray them as cartoonish characters satisfying appetites that are all too human – they kill, copulate and cannibalise to their heart's content. The Chapmans enlarged three of these cardboard dinosaurs for the installation *The Good, The Bad and The Ugly*, three gigantic dinosaurs in COR-TEN steel. The steel beasts have already appeared in various places around the world, including in Golders Hill Park in London and at the University of Warwick.

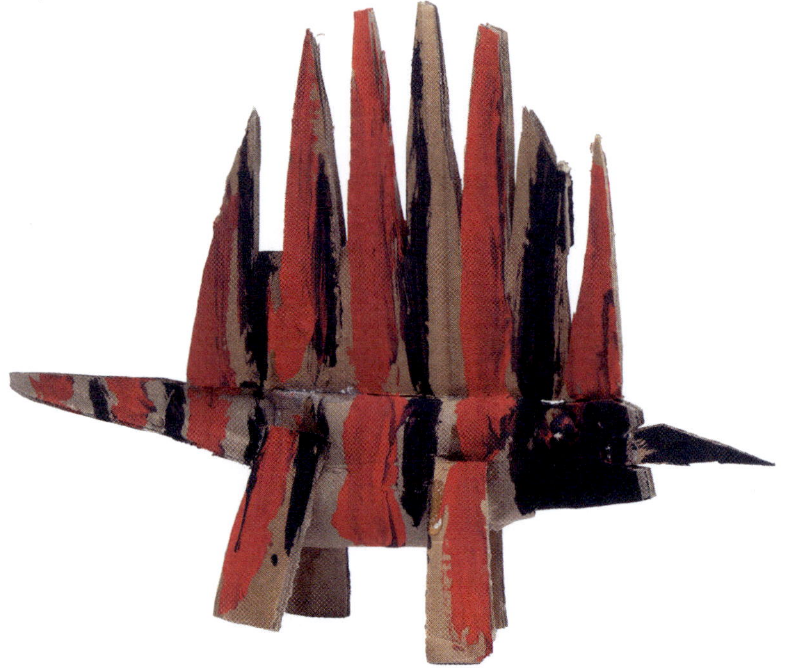

Hello Sixty Five Million Years BC, Jake and Dinos Chapman, 2004/2005.

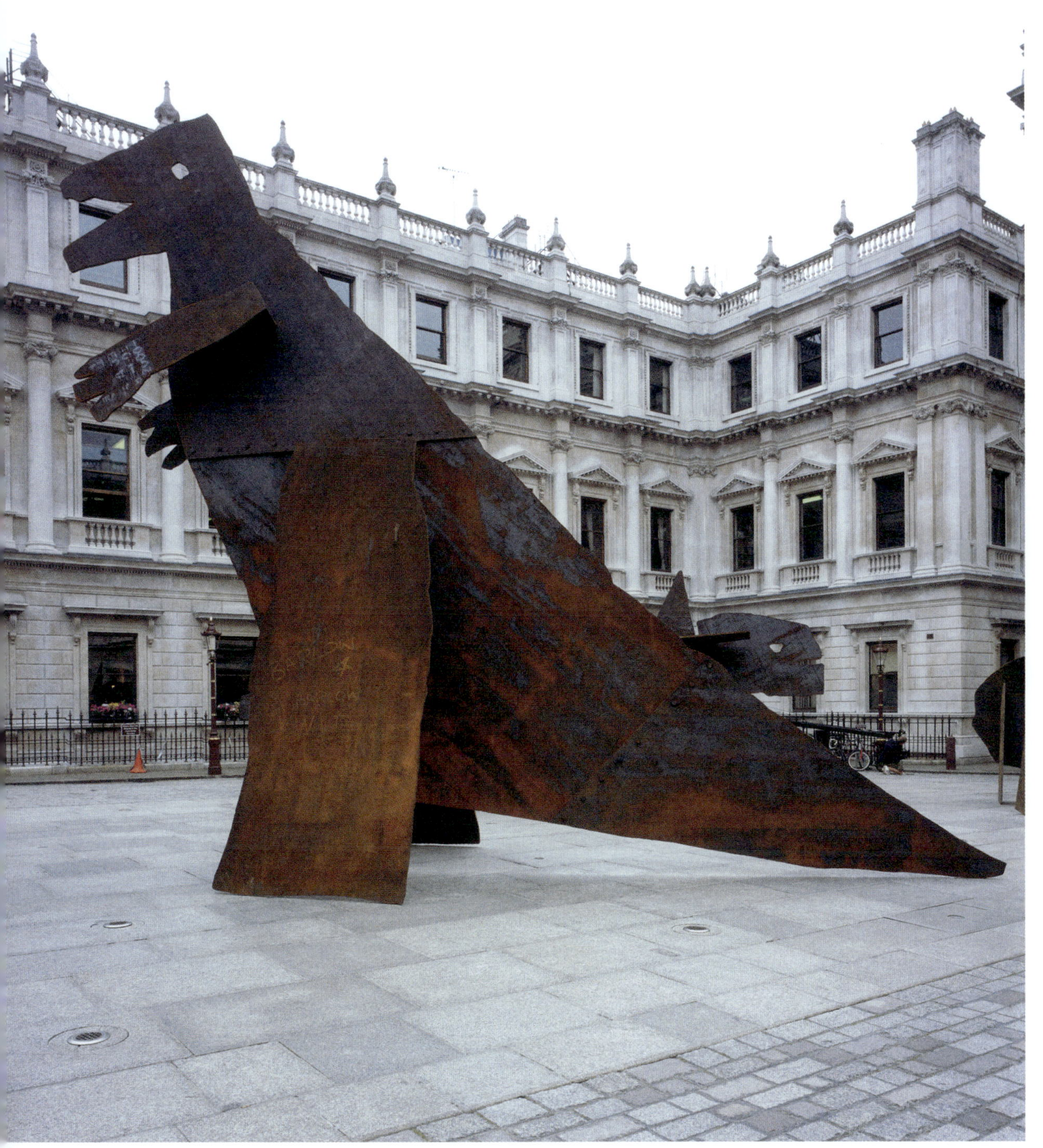

The Good, The Bad and The Ugly, Jake and Dinos Chapman, 2014.

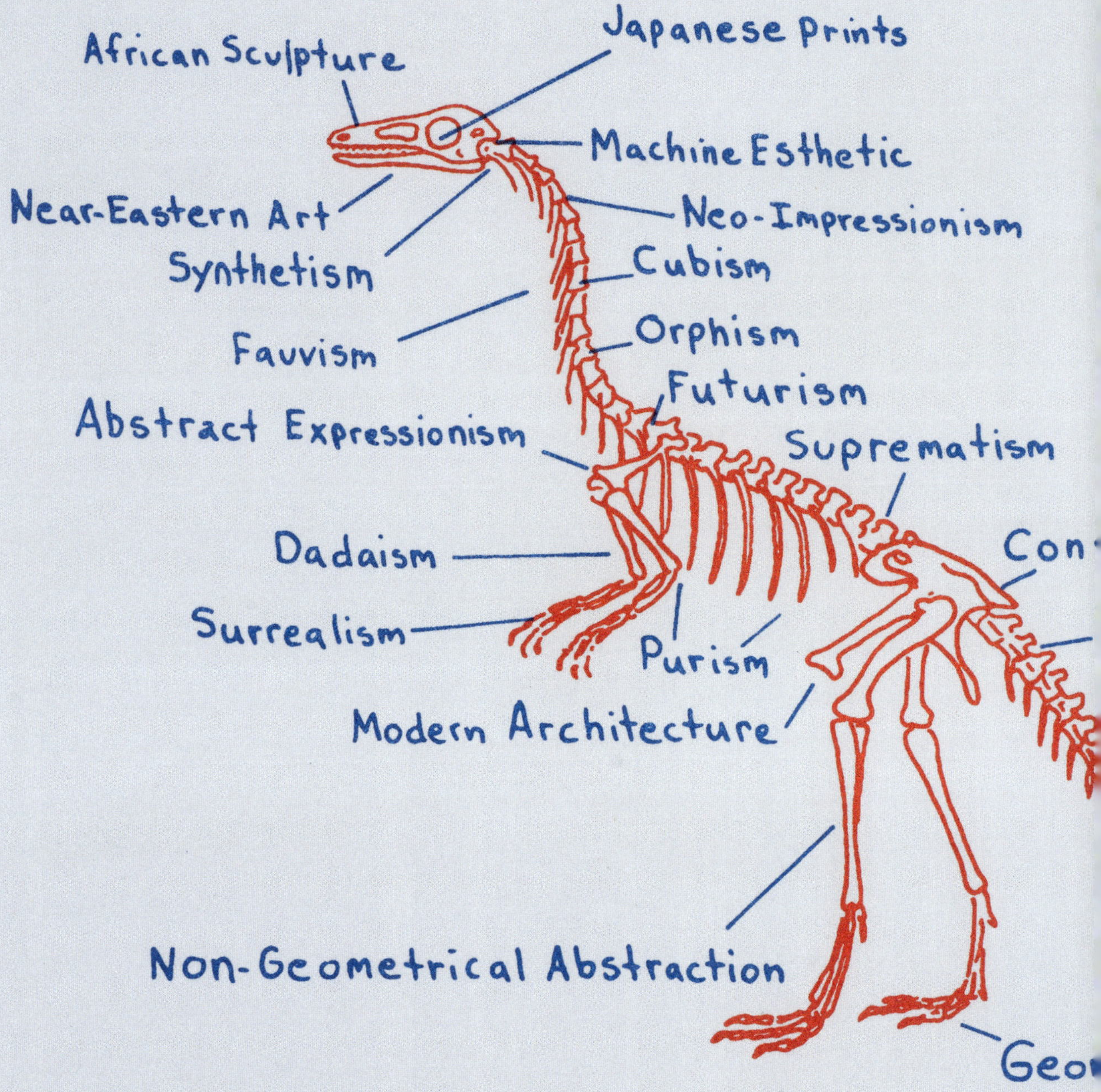

African Sculpture
Japanese Prints
Machine Esthetic
Near-Eastern Art
Neo-Impressionism
Synthetism
Cubism
Fauvism
Orphism
Futurism
Abstract Expressionism
Suprematism
Dadaism
Con
Surrealism
Purism
Modern Architecture
Non-Geometrical Abstraction
Geo

When Dinosaurs Ruled the Earth is a series of works by Mark Dion (b. 1961). In his oeuvre, the American contemporary artist questions how museums and institutions influence our understanding of the natural world and history. Given that museums present themselves as objective, authoritarian centres of knowledge, Dion often uses presentation techniques in his artworks that he borrows from the museum world. He amasses objects and curios, assembling them in display cases; his taxonomic installations often resemble cabinets of curiosity. Archaeology and palaeontology are central to his work. In 1999, he even organised a pseudo-archaeological expedition along the banks of the Thames in London. With a team of volunteers, he combed the river banks, discovering human bones, clay pipes, plastic shoes and a bottle containing a letter. These chance finds were presented in traditional display cabinets at Tate Modern to lend them 'scientific weight'.

On occasion, Dion playfully misemploys the 'seriousness' of scientific illustrations to reconstruct art history. The Smithsonian American Art Museum collection contains the lithograph *When Dinosaurs Ruled the Earth* from his *World in a Box* suite in which Dion uses a drawing of an *Iguanodon* skeleton to situate art historical movements. In their illustrations, palaeontologists often misrepresented dinosaurs because knowledge of dinosaur morphology was still very inexact, which is why those 'scientific sketches' should always be seen in their time and context. According to Dion, you can prove anything with a drawing. Although, of course, using a dinosaur skeleton to map the course of art history is impossible. Dion uses scientific modes of classification to peddle nonsense. Palaeontology is not art, but sometimes science – like art – also balances fact and fiction.

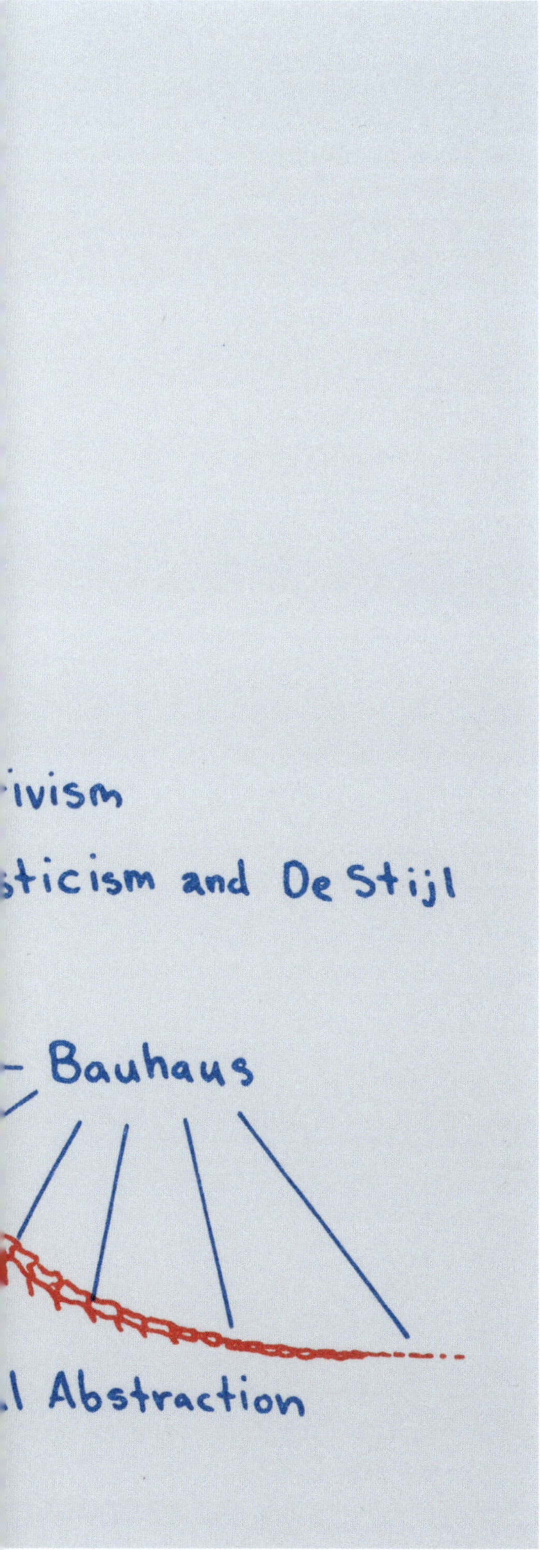

———

<
When Dinosaurs Ruled the Earth, Mark Dion, 2020.

RENDERED DINOFOSSILS

Multimedia artist Kerstin Brätsch (b. 1979) gives fossils a contemporary interpretation. Petrified remains are created in a process that can take millions of years. Brätsch makes synthetic marble, but uses time-honoured craft techniques and pigments. In this way, she 'fossilises' time, (art) history and her personal history in her artificial 'stone'. In a recent installation in the MoMa café, Fossil Psychics for Christa, she created a trompe-l'oeil effect with the black marble that was used to build the bar, with wallpaper. The faux marble wall utilises 3D renderings of dinosaurs. The final piece, entitled Dino Runes [Towards an Alphabet], is an anachronistic mixed media work that allows natural stone and fauna millions of years old to time travel, thanks to new digital technology. Just as fossils are sometimes difficult to decipher geologically and paleontologically, her work is a new language whose alphabet the viewer has yet to learn.

———

PSYCHEDELIC *T. REX*

Liz Markus (b. 1967) is certainly not one of those paleoartists who strive for scientific accuracy in their depictions of dinosaurs. Markus's paintings of tyrannosaurs are very intuitive, hazy and atmospheric, loosely based on romanticised images of dinosaurs from popular media or films. Just as children's toys rarely accurately reproduce dinosaur skin, Markus chooses her colours purely on gut feeling. The result is naive, psychedelic tyrannosaurs, sometimes shown alone, with a radiant halo of acrylic paint, sometimes locked in a duel with Frankenstein's monster or some other creature.

———

>

Giant Green T. Rex with Auras, Liz Markus, 2020.

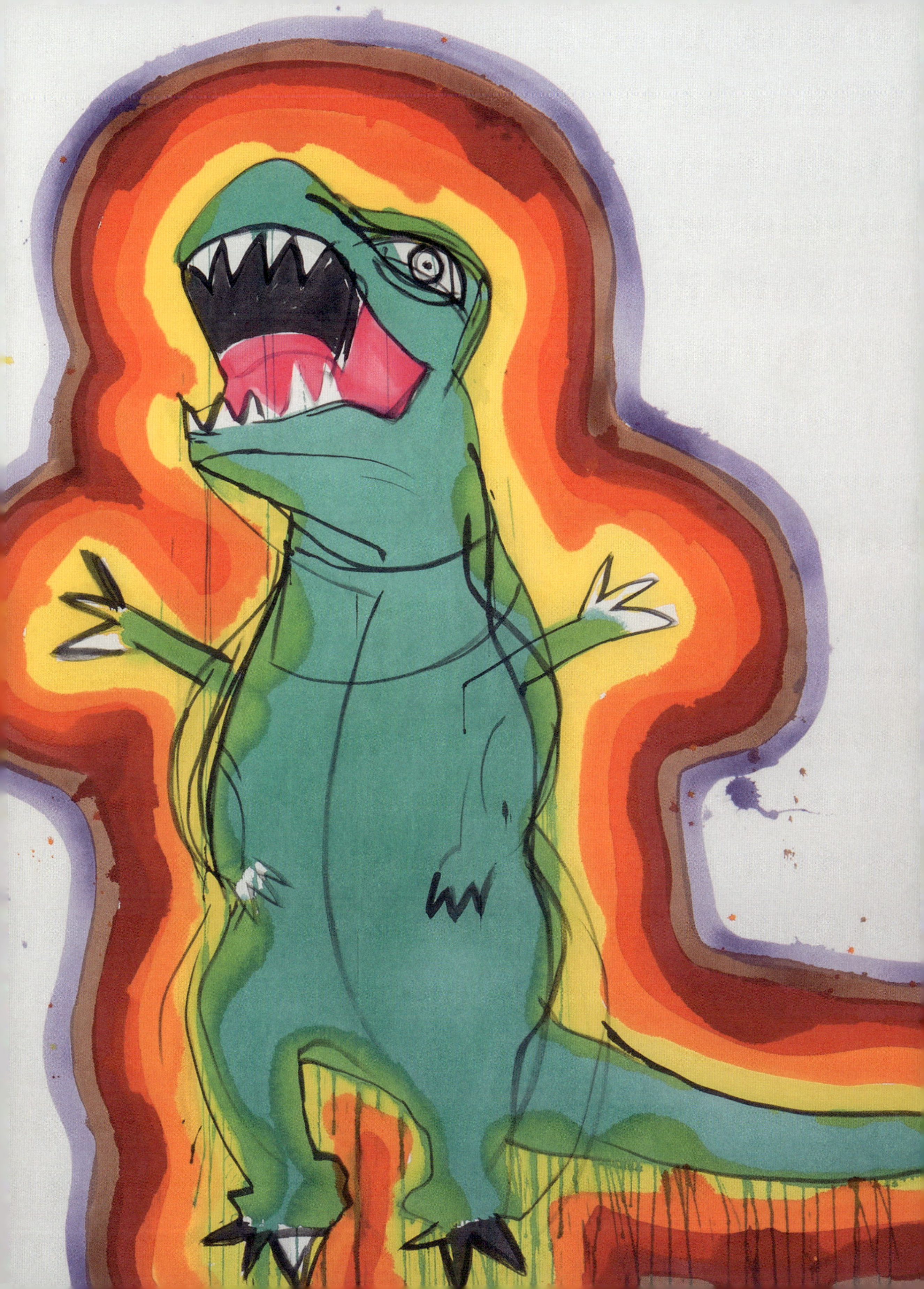

DINOSAUR WITH A MICROCHIP

The Belgian artist Panamarenko (born Henri Van Herwegen, 1940–2019) is best known for his imposing airships, jetpacks, pedal- and electric-powered propeller kites, speedboats and flying saucers. Fascinated by science and nature, he began experimenting with all kinds of flying objects in the 1960s and 1970s and took inspiration for some of his works from the vibrating mechanism of insect wings. He also became increasingly interested in prehistoric primeval birds and between 1989 and 1991 built the first series of a Archaeopteryxes, which he dubbed the 'Flying Chicken'. The design is based on the 'primordial bird' *Archaeopteryx lithographica* first found in the lithographic limestone of southern Germany (see Chapter 2); by integrating photocells into the wings, and relays, servo motors and microchips into the abdomen, he enabled the robot to walk about on its own. The microchips caused technical control difficulties, and in his second generation of robotic chickens he gradually replaced the electronics with a fully mechanical system. This was followed in 2001 by the *Persis Clambatta* (named after the actress Persis Khambatta) and in 2004 by *De Vogelmarkt* (The Bird Market), an installation of three tables for three archaeopteryxes to walk on. In the Middelheim Museum in Antwerp you can see an archaeopteryx from 1993 that jumps and flaps when you approach the sculpture.

———

>

Persis Clambata, Panamarenko, 2001.

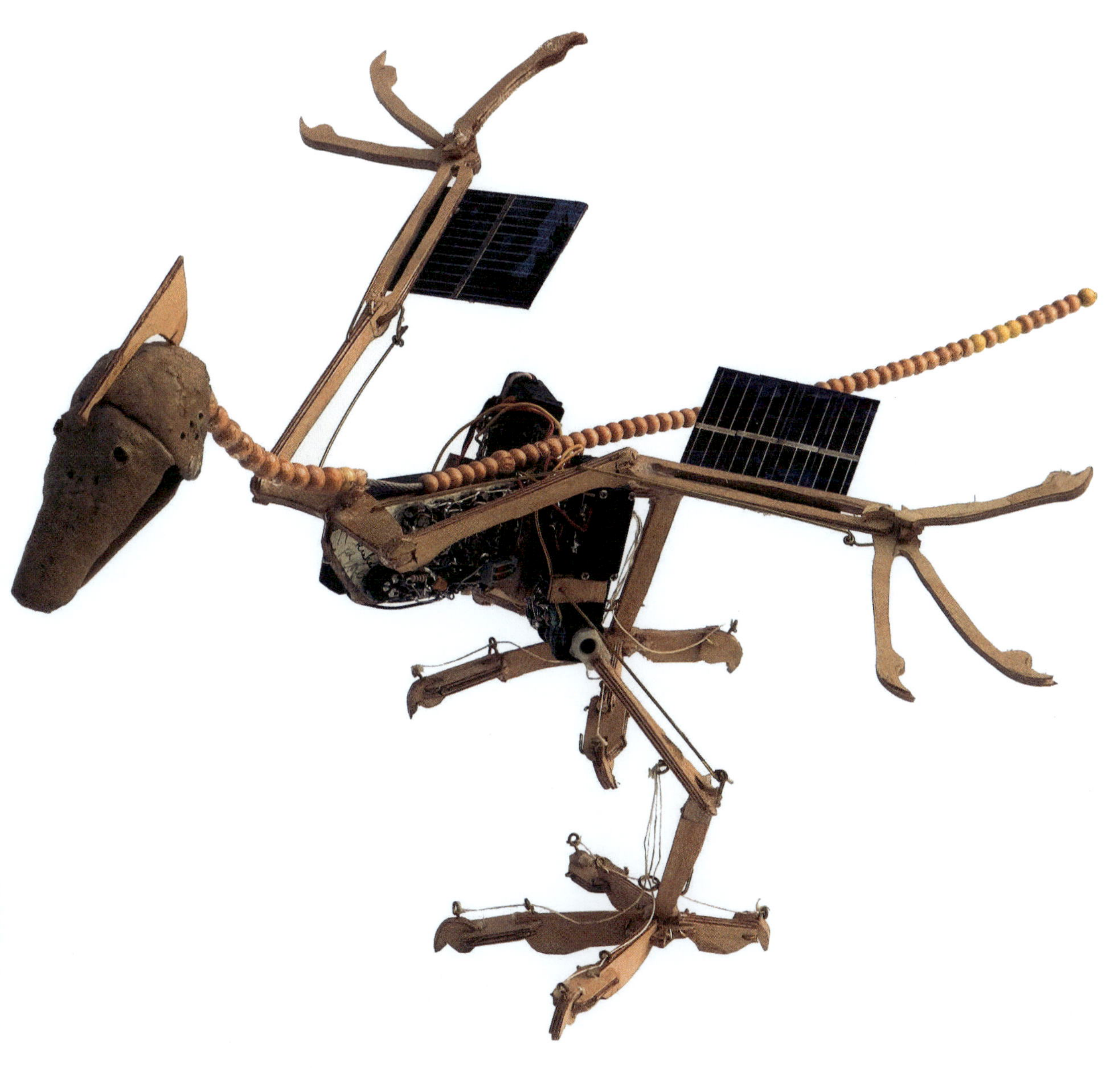

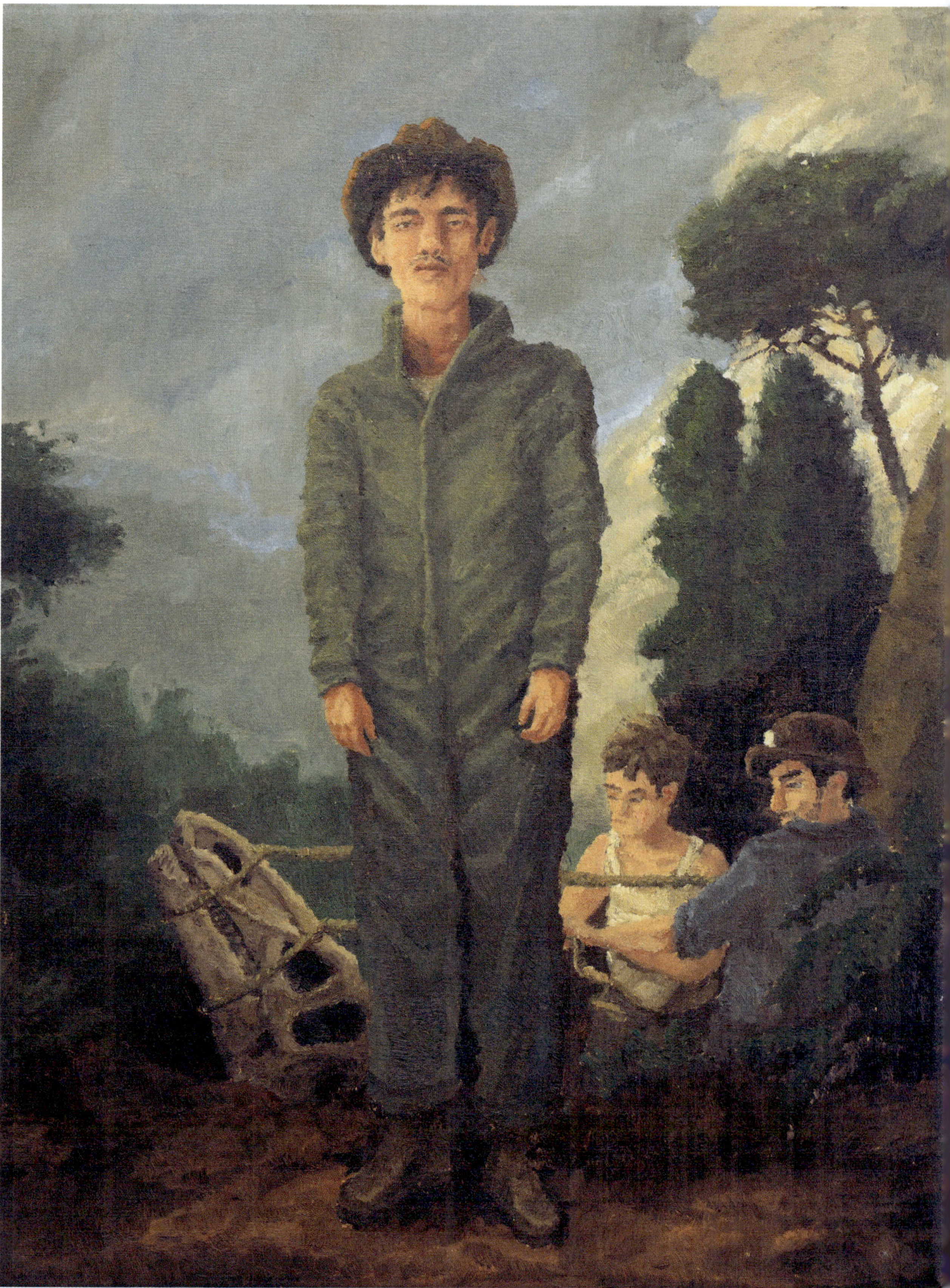

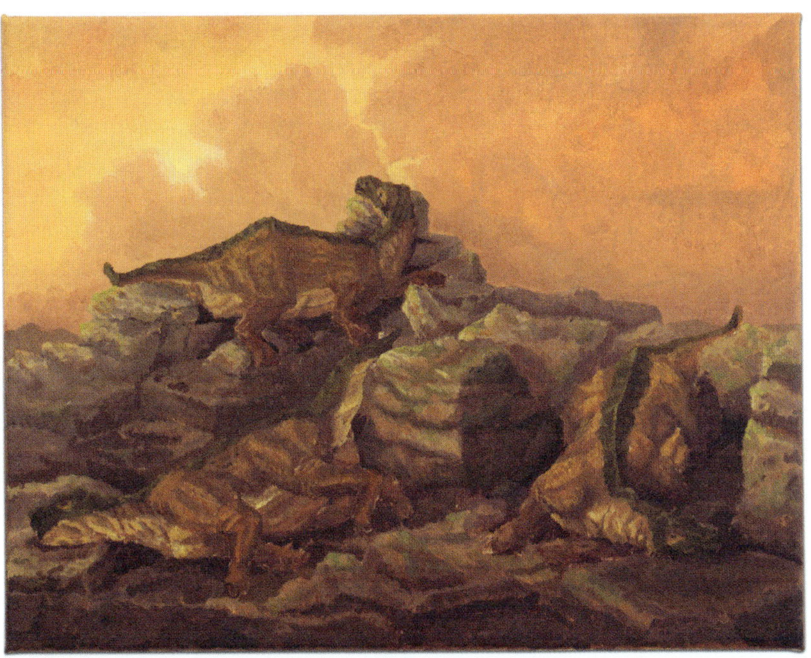

NOSTALGIA FOR BERNISSART

For his recent exhibition *Blackbird on a Shoulder* at the Sofie Van de Velde gallery in Antwerp, the young Belgian artist Felix De Clercq created a series of paintings and sculptures inspired by 19th-century palaeontology, more specifically by the historical discovery of some 30 *Iguanodon* skeletons in the Bernissart coal mine (see Chapter 2). The discovery immediately put Belgium on the worldwide palaeontological map. De Clercq collected anecdotes and old documents related to the finds. Those sources of inspiration fossilised into a series of intimate paintings and sculptures, in which he combines romance and nostalgia with a boyish sense of adventure. De Clercq even found a 19th-century miner's helmet that he turned into a ready-made. One painting takes inspiration from a historical photograph from 1882 and shows the team of palaeontologists from the Belgian Institute of Natural Sciences assembling the first complete *Iguanodon* skeleton. Due to the creature's size, they needed to work in a high-ceilinged space and were unable to mount the remains in the Brussels museum. They moved to the 16th-century St George's Chapel of the Royal Library in Brussels, where the photograph was taken. The Belgian artist Léon Becker also painted the same scene in 1884. That work is now part of the permanent collection of the Belgian Institute of Natural Sciences.

---

^

Sisters, Felix De Clercq, 2021.

<

Luca, Felix De Clercq, 2021.

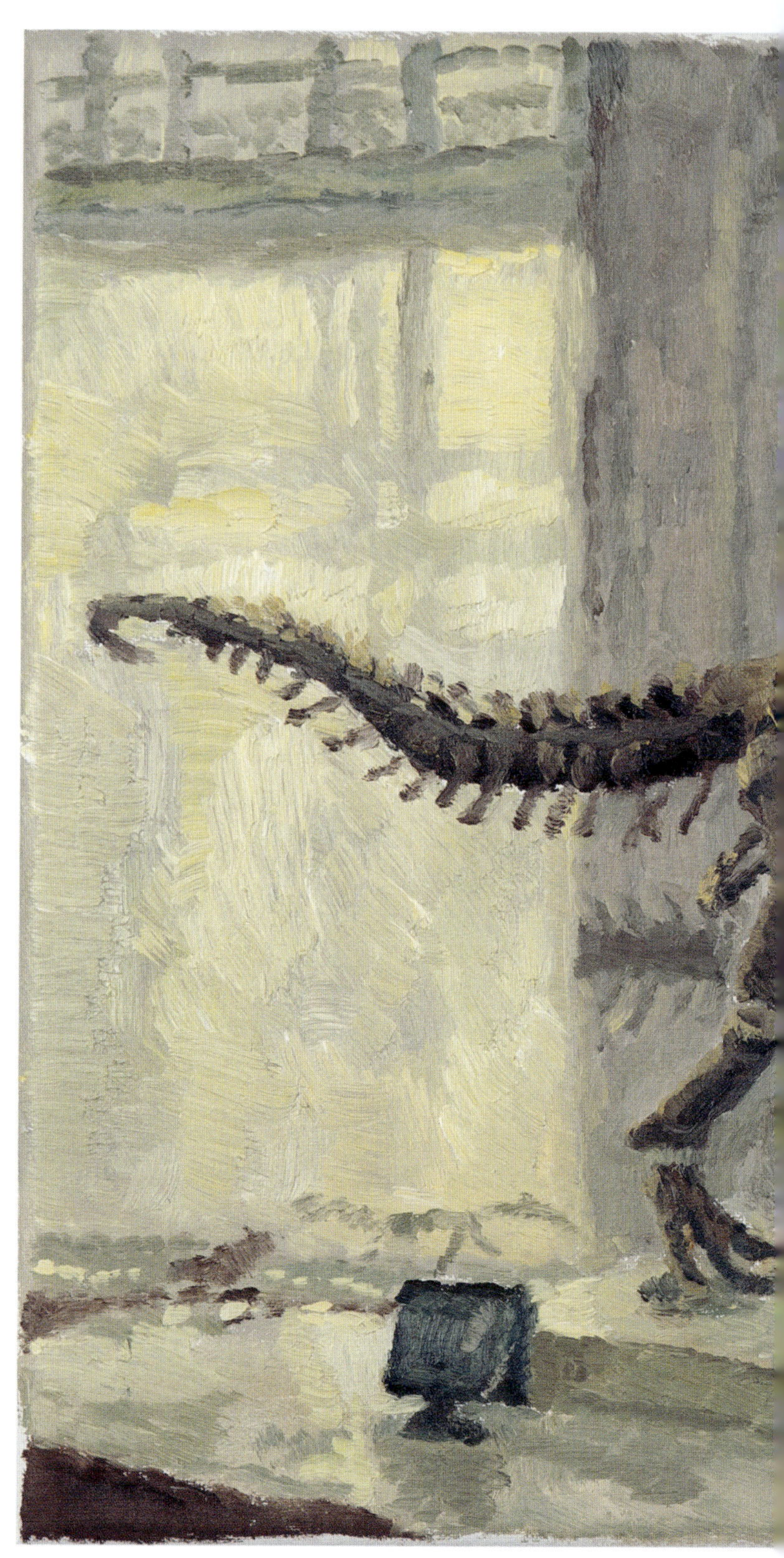

Arkhane Allosaurus sold at an auction at
the Eiffel Tower, Felix De Clercq, 2021.

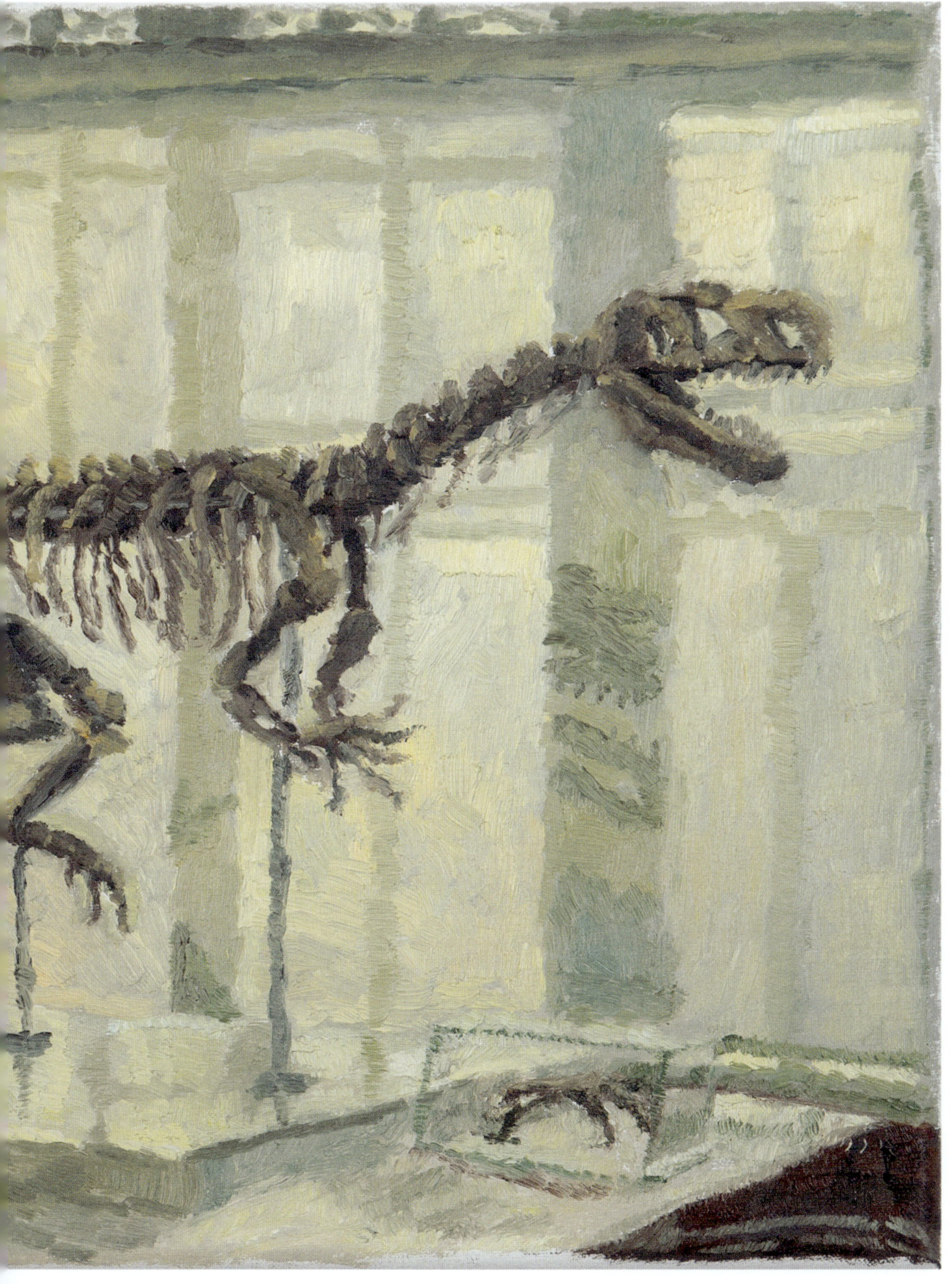

WHAT IS THE CONNECTION BETWEEN ARTHUR CONAN DOYLE AND DINOSAURS?

In 1852, Charles Dickens became the first novelist to include a dinosaur in a work of fiction. In *Voyage au centre de la terre* (1864), the French writer Jules Verne talked of an extinct creature that featured the traits of a horse, hippo, rhinoceros and camel, which God had hastily flung together in the first hours of creation. But it was not until 1912 that the first real dinosaur novel saw the light of day. This was the year in which Sir Arthur Conan Doyle, the author of the Sherlock Holmes detective stories, wrote *The Lost World*. In the novel, Professor Challenger discovers a primordial strip of jungle on a plateau in Mexico, populated by prehistoric creatures. They find the dinosaurs, although members of Challenger's team narrowly escape death. The novel ends with the team making it back to London safely, bringing with them a baby *Petrodactylus*, a flying reptile from the Late Jurassic Period. The book lives on in science. A species, discovered in present-day Brazil in 1996, has been named after its protagonist: *Irritator challengeri* (a species of spinosaurid).

Conan Doyle's *The Lost World* was filmed in 1925, with the author playing a supporting role. The film became the very first in-flight movie to be shown on the Imperial Airways London–Paris flight. The film was very successful due to its revolutionary stop-motion effects. These were devised by Willis O'Brien, the man who would also work on the first *King Kong* film in 1933. The dinosaur models used in the film were based on Charles Knight's dinosaur paintings, which he made for the Museum of Natural History in New York. The models were eventually donated to the Museum of Arts and Sciences in Los Angeles. Over the years, the rubber models began to degrade and crumble, and were finally lost when a new wing to the museum was renovated.

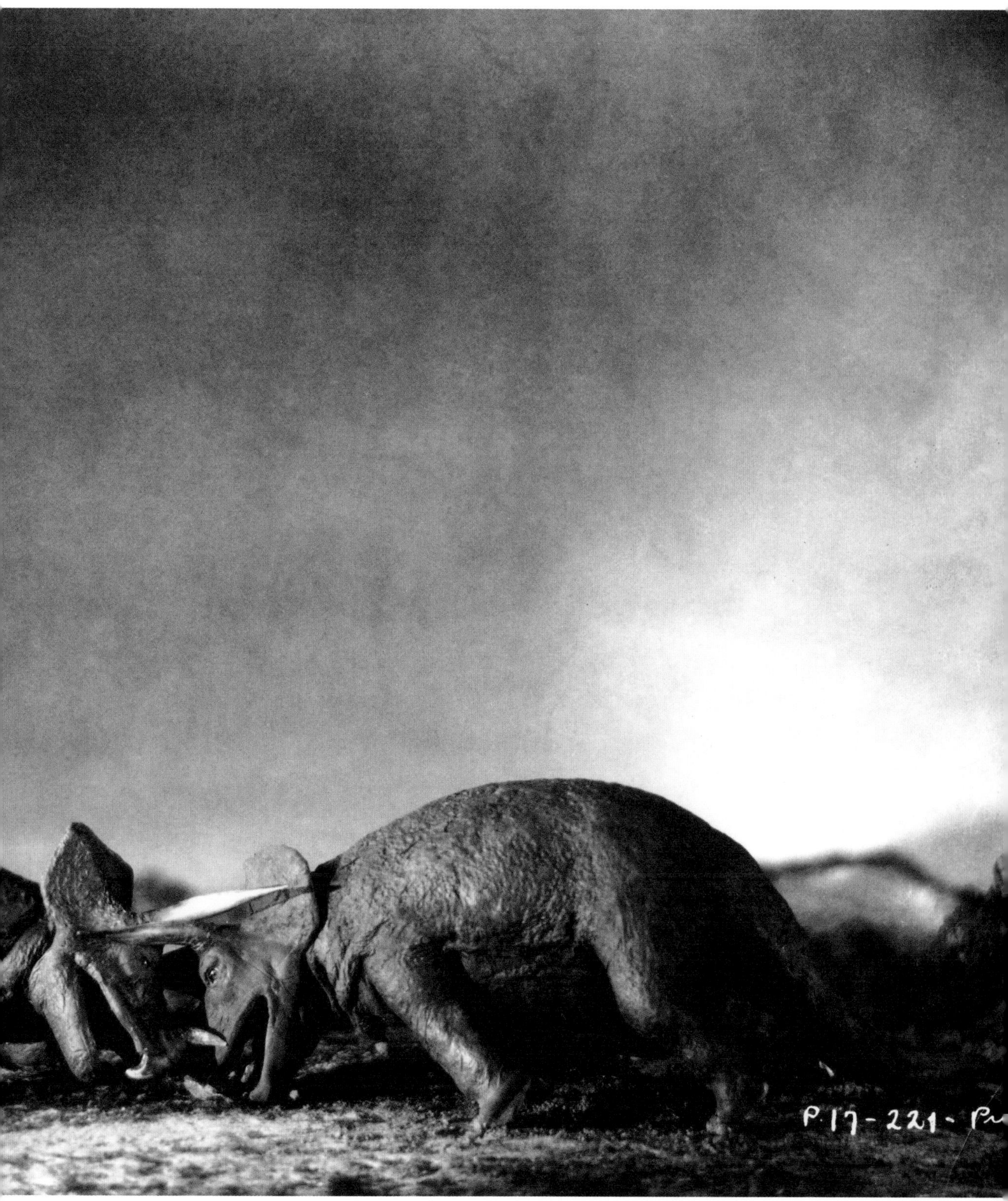

Stop-motion scene from the 1935 film of Arthur Conan Doyle's *The Lost World*.

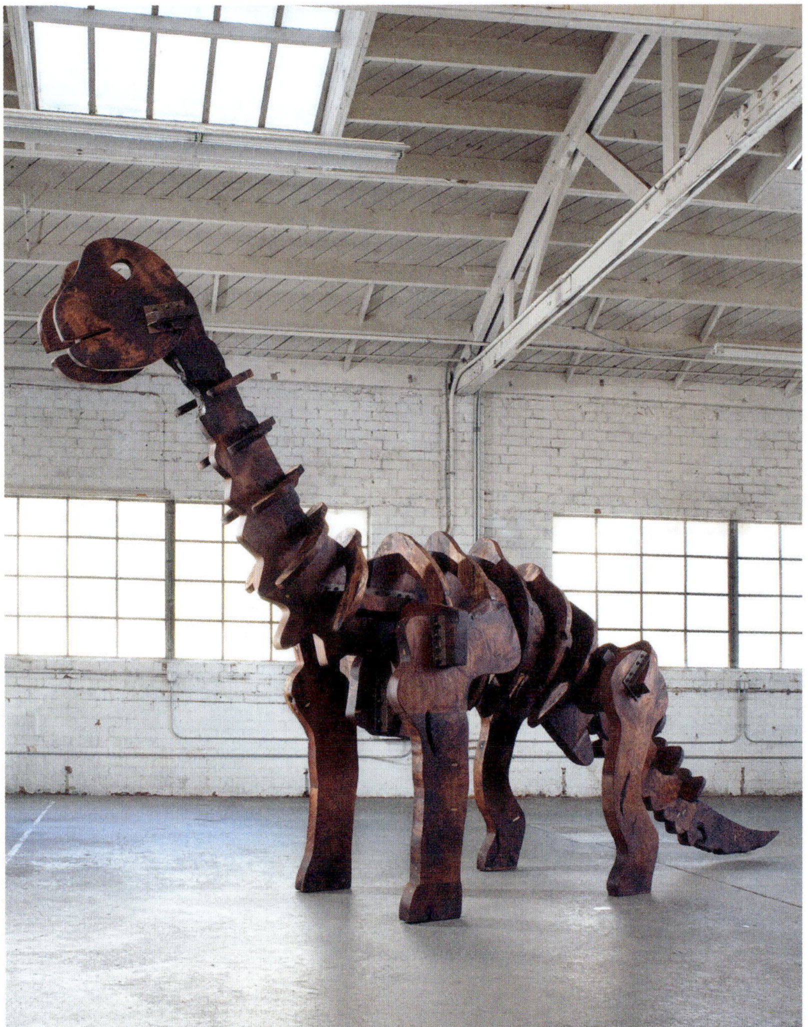

Baby Dinosaur (Apatosaurus), Matt Johnson, 2013.

GIANT BABY

Baby Dinosaur (Apatosaurus) is a giant sculpture by Matt Johnson (1978). In 2013, the American artist recreated a full-size skeleton of a baby *Apatosaurus*. Morphologically, the sculpture is far from accurate. The artwork looks more like a magnified version of one of those wooden dinosaur skeletons designed as a children's puzzle. For the sculpture, Matt Johnson used steel and sequoia wood, a species of tree that also existed in the time of the dinosaurs.

After stuffed animals, Darwin, Sinke & Van Tongeren
are now working with dinosaur skeletons.

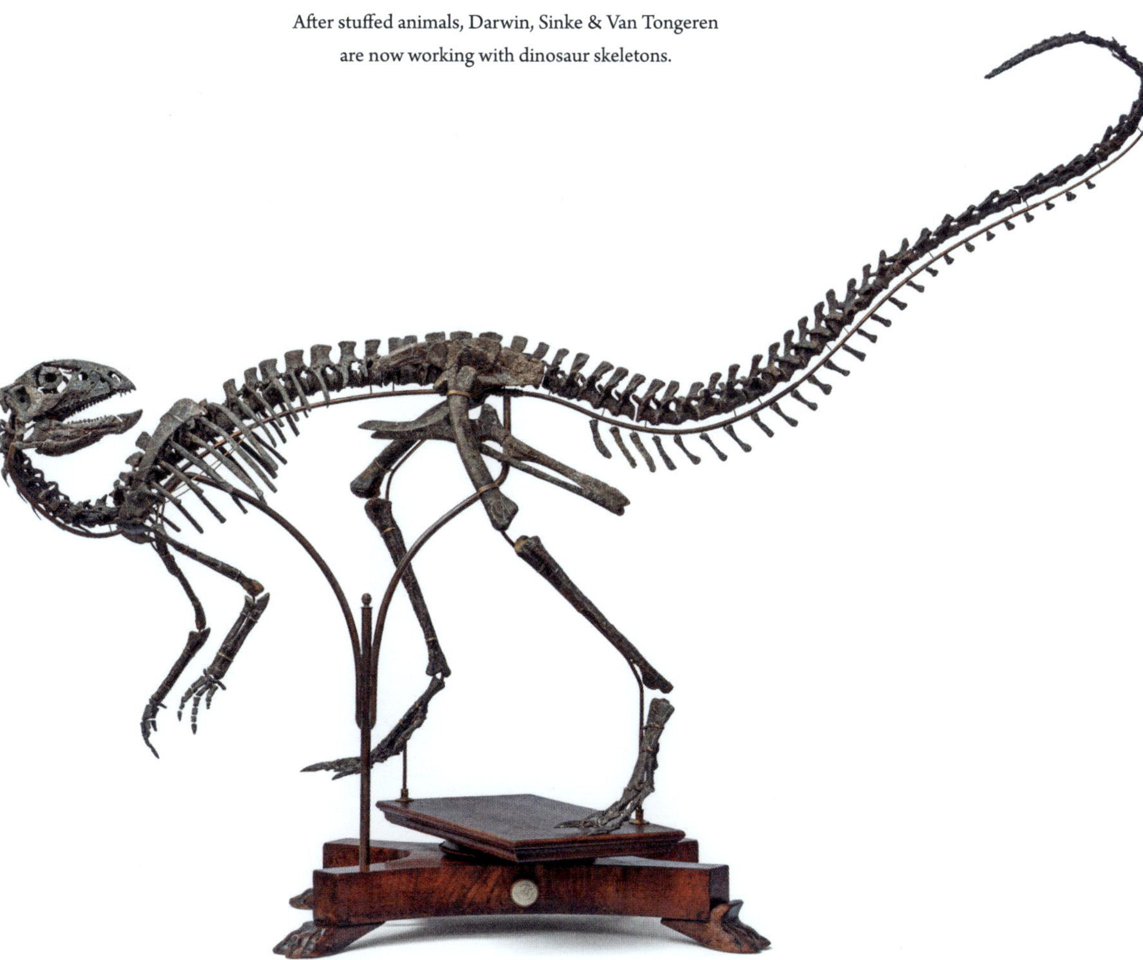

TAXIDERMIED DINOSAURS

In recent years, Darwin, Sinke & Van Tongeren have amazed the art world
with their breathtaking taxidermic compositions. Their practice not only em-
braces mounting stuffed animals in spectacular, lifelike poses, but also illu-
minates the techniques of the craft. The duo has also photographed animals
in their milky tanning bath, creating haunting images of a creature hovering
between life and death. The Dutch artists are currently collaborating with a
palaeontologist from Italy to create a new series of sculptures based on dinosaur bones.

———

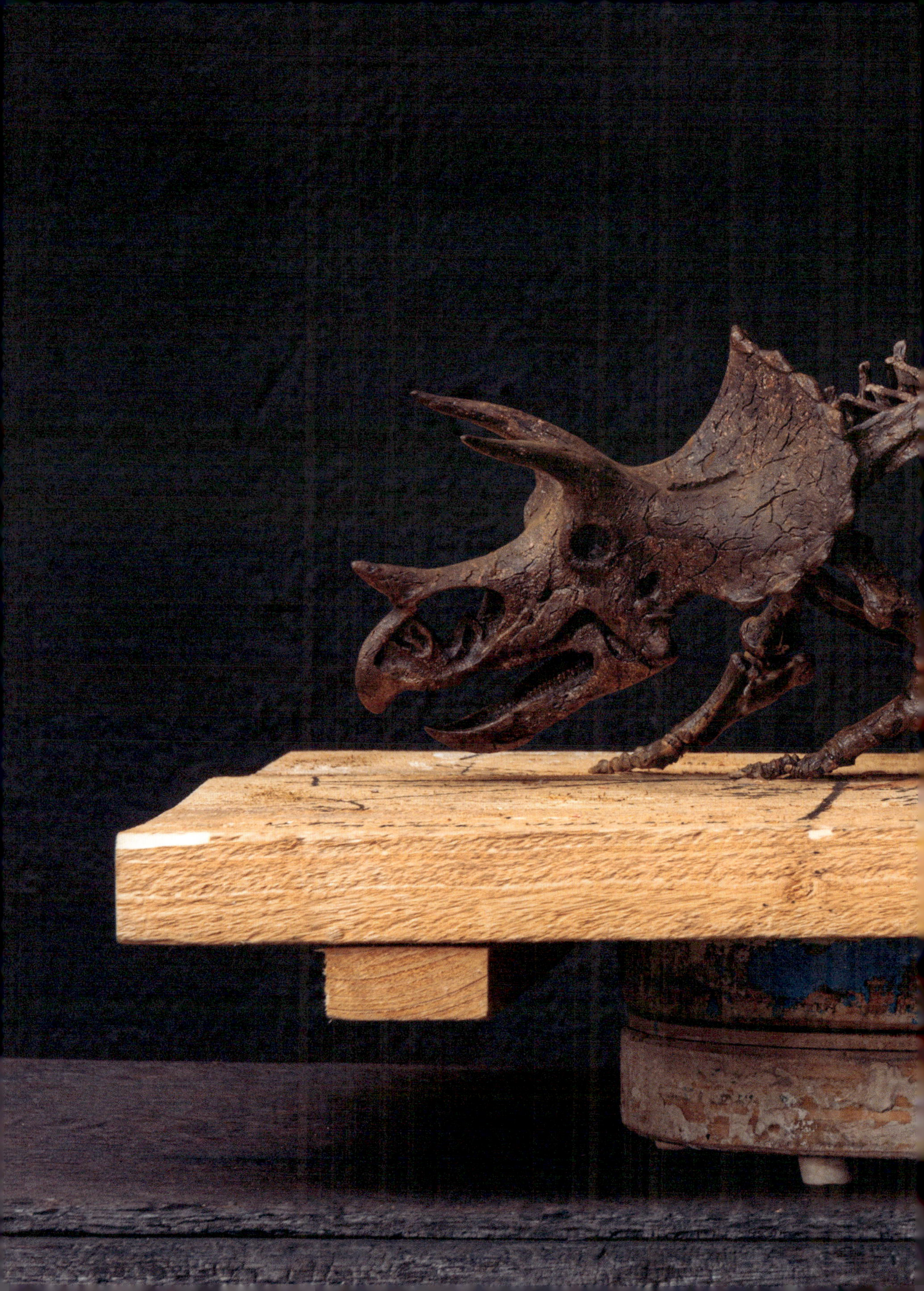

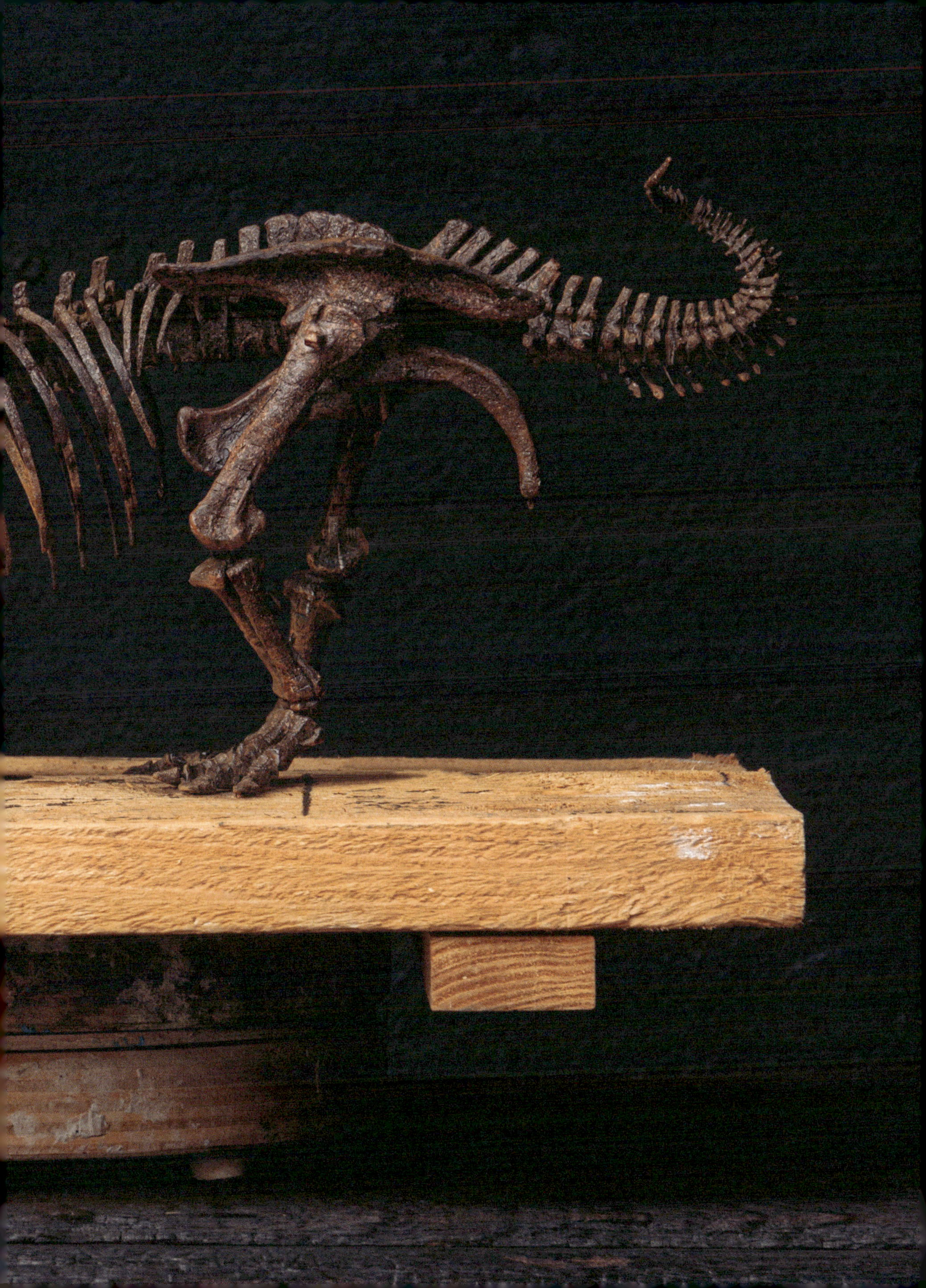

WHAT IS THE CONNECTION BETWEEN CHARLES DICKENS AND DINOSAURS?

When did dinosaurs first appear in literary fiction? Probably in the novels of Charles Dickens (1812–1870). The author of *Oliver Twist* and *A Christmas Carol* was good friends with British palaeontologist Richard Owen (1804–1892), the man who coined the term 'dinosaur' in 1842, after the recent discoveries of *Iguanodon* and *Hylaeosaurus* among others. That friendship left several traces in Dickens's oeuvre, beginning with *Household Words*, the magazine Dickens published weekly between 1850 and 1859. In 1851 – the year of the first World Expo, or Great Exhibition, in London – he published a short story by Henry Morley. In it, Morley fantasised about a place in South America where dinosaurs still existed, and attempted a description of a *Megalosaurus,* the first dinosaur species ever described in 1824 and one of the few known at the time. 'Here's a land reptile, before which we take the liberty of running. His teeth look too decidedly carnivorous. A sort of crocodile, thirty feet long, with a big body, mounted on high thick legs, is not likely to be friendly with our legs and bodies. Megalosaurus is his name, and, doubtless greedy is his nature.'

The following year, Dickens also featured a *Megalosaurus* in his 1852–1853 novel *Bleak House.* He even begins his book with it: 'London. Michaelmas Term lately over, and the Lord Chancellor sitting in Lincoln's Inn Hall. Implacable November weather. As much mud in the streets as if the waters had but newly retired from the face of the earth, and it would not be wonderful to meet a *Megalosaurus,* forty feet long or so, waddling like an elephantine lizard up Holborn-Hill.' Dickens is said to have borrowed the term 'elephantine lizard' from his friend Richard Owen. The timing of *Bleak House* is unusual, because it was not until 1854 that Owen published the first drawing of an entire *Megalosaurus* – two years after the publication of *Bleak House* – although we can assume that Owen had already shown the sketch to Dickens for inspiration for his book. Nor would Dickens have been able to draw inspiration for his book from the *Megalosaurus* in the Dinosaur Sculpture Park near the Crystal Palace, because it did not open until 1854. But, because Owen was also involved in that project, Dickens could very probably have known of it.

Owen himself plays a walk-on part in Dickens's last completed book, published in 1864, *Our Mutual Friend.* In the second chapter, he gives an account of a dinner party. One of the dinner guests is a certain Mrs Posdnap, 'a fine woman for Professor Owen, with a quantity of bone, neck and nostrils like a rocking-horse'. A year later, in February 1865, Dickens published an – admittedly belated – eulogy for Mary Anning (1799–1847) in his magazine *All the Year Round* (see Chapter 2). He thought that the British palaeontologist deserved more recognition than she had received so far. 'But Mary Anning was something more than a mere village celebrity. She acquired, if not an English, certainly a European reputation. Professor Owen thought so highly of her usefulness, that he moved the authorities of the British Museum to grant her a pension of forty pounds a year, which she enjoyed for some little time before her early death,' we read in Dickens's magazine. Anning died from breast cancer at the age of 47. When she was diagnosed, the Geological Society of London organised a fund-raising drive – out of respect – to pay for her treatment. However, as a woman, she was not permitted to become a member of the society.

Portrait of Charles Dickens by Johann Jacob Weber.

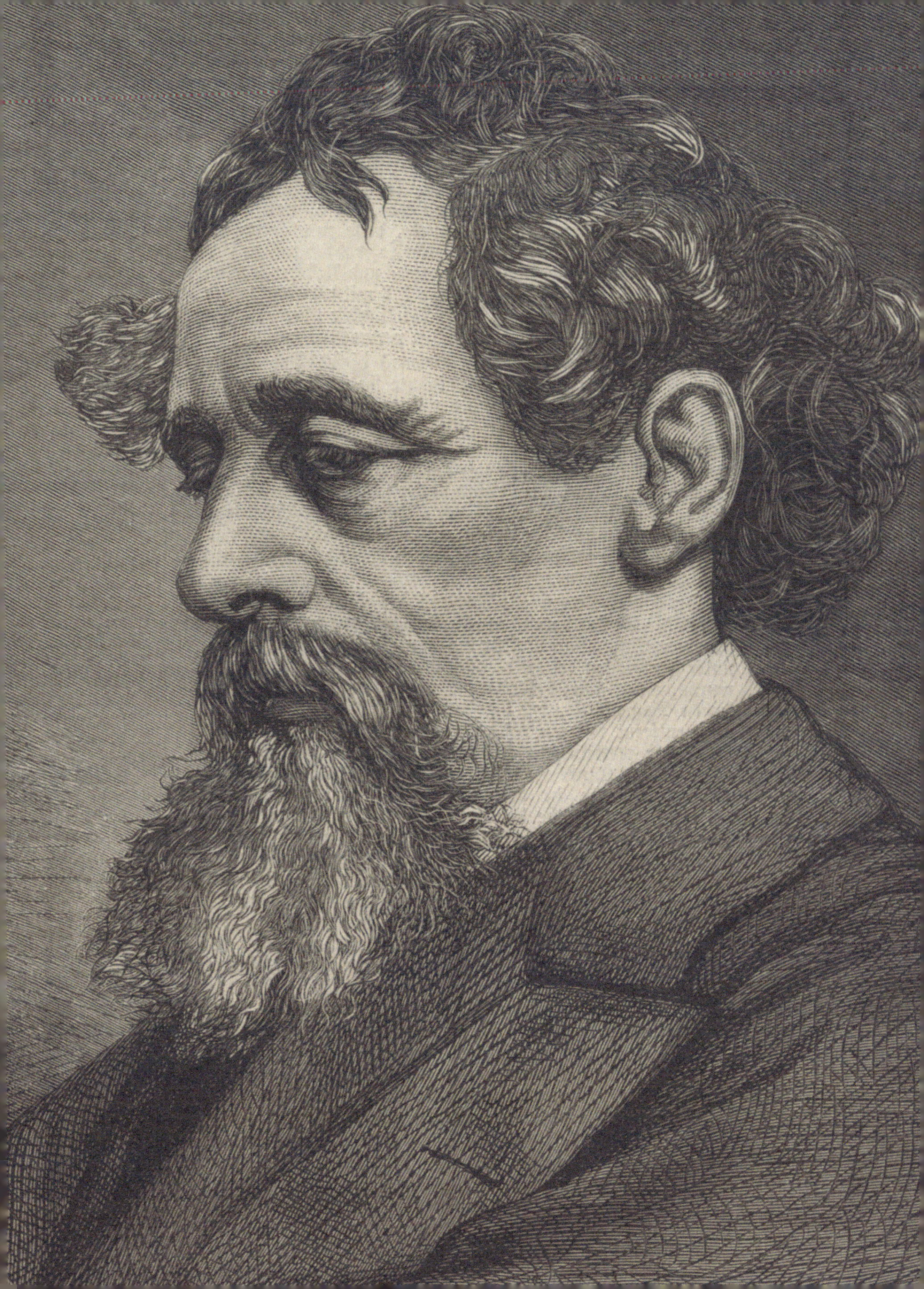

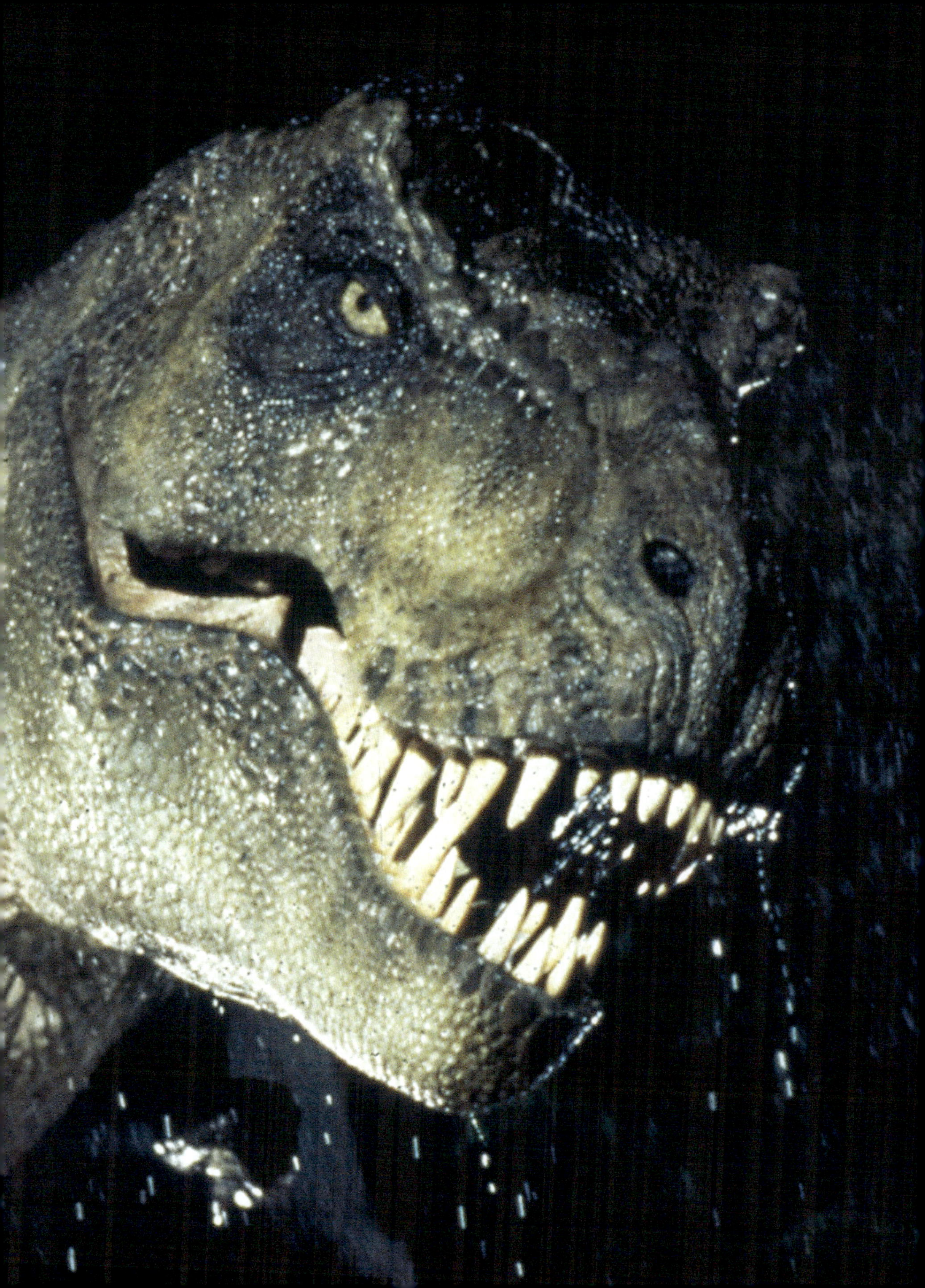

CHAPTER 5

DINOSAURS OF THE FUTURE

CAN JURASSIC PARK COME TRUE?

'Your scientists were

so preoccupied with whether they could,

they didn't stop to think if they should.'

Dr Ian Malcolm in Jurassic Park, 1993

The *Jurassic Park* film and its many sequels continue to fascinate new generations of dinosaur fans. But has science now reached the point where the film's futuristic scenario can come true? Could we ever bring dinosaurs back to life using cloned DNA extracted from dinosaur blood consumed by an ancient, fossilised mosquito? Does this make scientific sense? How close is science to resurrecting dinosaurs? Should palaeontologists even try to bring these extinct animals back to life?

Some scientists secretly dream that, one day, they'll be able to resurrect dinosaurs by using DNA or genetic engineering. Scientists have been experimenting for years, but things seem to have reached a deadlock. Which might be a good thing. After all, the film *Jurassic Park* – and Michael Crichton's 1990 novel on which it was based – revolve around the risks of genetic engineering. 'Don't you see the danger, John, inherent in what you're doing here? Genetic power is the most awesome force the planet's ever seen, but you wield it like a kid that's found his dad's gun,' says Dr Ian Malcolm in the film. 'Your scientists were so preoccupied with whether they could, they didn't stop to think if they should.'

<

Film still from Steven Spielberg's *Jurassic Park* (1993).

For films like *Jurassic Park* (and its sequels) and for documentaries such as *Walking with Dinosaurs*, the animators worked with a team of palaeontologists. With their input, they were able to create an accurate portrayal of the prehistoric animals, both in appearance and in behaviour. The risk – inherent in paleoart – is that science quickly catches up with this celluloid image. 'In the first *Jurassic Park* film in 1993, *Tyrannosaurus* still had a reptilian skin; we now know they had feathers,' writes Michael J. Benton. But while that evidence has been around since 2004, the *T. rex* in the 2015 and 2020 film sequels is featherless. 'We want the dinosaurs to look scary, and that means big teeth and scaly skin,' one of the producers must have said. 'A feathery *T. rex* would just look like an overgrown chicken'.

Benton, like British palaeontologist Mark Witton, also worked on dinosaur film productions or documentaries, such as the late-1990s BBC series *Walking with Dinosaurs*. The scientists advised the team based on the then-current palaeontological knowledge about how the dinosaurs looked, moved and fed. Looking at it today, it's easy to spot numerous inaccuracies, all of them due to new scientific insights, although the filmmakers also sometimes based choices on dramatic effect, or the story's function. 'I worked on the pterosaurs,' says Witton. 'I was very happy with the animations we made: in my opinion, they were the best pterosaurs ever. But the 3D team changed the faces afterwards to make them more recognisable and graphic. Which came as a bit of a shock when it was shown on television.'

CAN YOU REALLY MAKE DINOSAURS
FROM MOSQUITOES?

Let's start with the premise of the first *Jurassic Park* from 1993. In that film, scientists experiment with blood that prehistoric mosquitoes sucked from a dinosaur. The DNA has been preserved because, shortly after biting the dinosaur, the mosquitoes landed on a tree and became trapped in its resin. This resin fossilised and eventually became amber. Insects caught in this amber would also turn into fossils. This means that they would also undergo certain processes of decay. Consequently, if the mosquito had consumed dinosaur blood and some hadn't yet been broken down by its digestive system, whatever blood remained would have deteriorated during the initial stages of fossilisation. Scientists have found excellently preserved insect muscle tissue in amber, but the digestive system is usually the first part of the body to degrade. So if you were to take a DNA sample from these fossilised mosquitoes, you would be cloning mosquitoes, and who would benefit from that?

———

COULD WE CLONE A MAMMOTH OR A DINOSAUR?

In 2021, scientists claimed to have the necessary technology and funding to resurrect the woolly mammoth. Admittedly, a mammoth isn't a dinosaur, but it's moving in the right prehistoric direction. For the record, the plan was not about cloning a mammoth. After all, that's impossible – even the available DNA from the best-preserved mammoth is far too degraded. What scientists wanted to do was to create a mammoth–elephant hybrid that, too look at, would be indistinguishable from the extinct woolly mammoth. The research is still ongoing and funding isn't guaranteed – but is it a stepping stone to a cloned dinosaur? Is it possible that, one day, we'll be able to clone dinosaurs with ancient DNA? Even if only in hybrid form? Probably not. If 10,000-year-old mammoth DNA is too damaged, there's not much hope of finding intact dinosaur DNA that's 66 million years old.

———

>

Steven Spielberg and baby *Stegosaurus.*

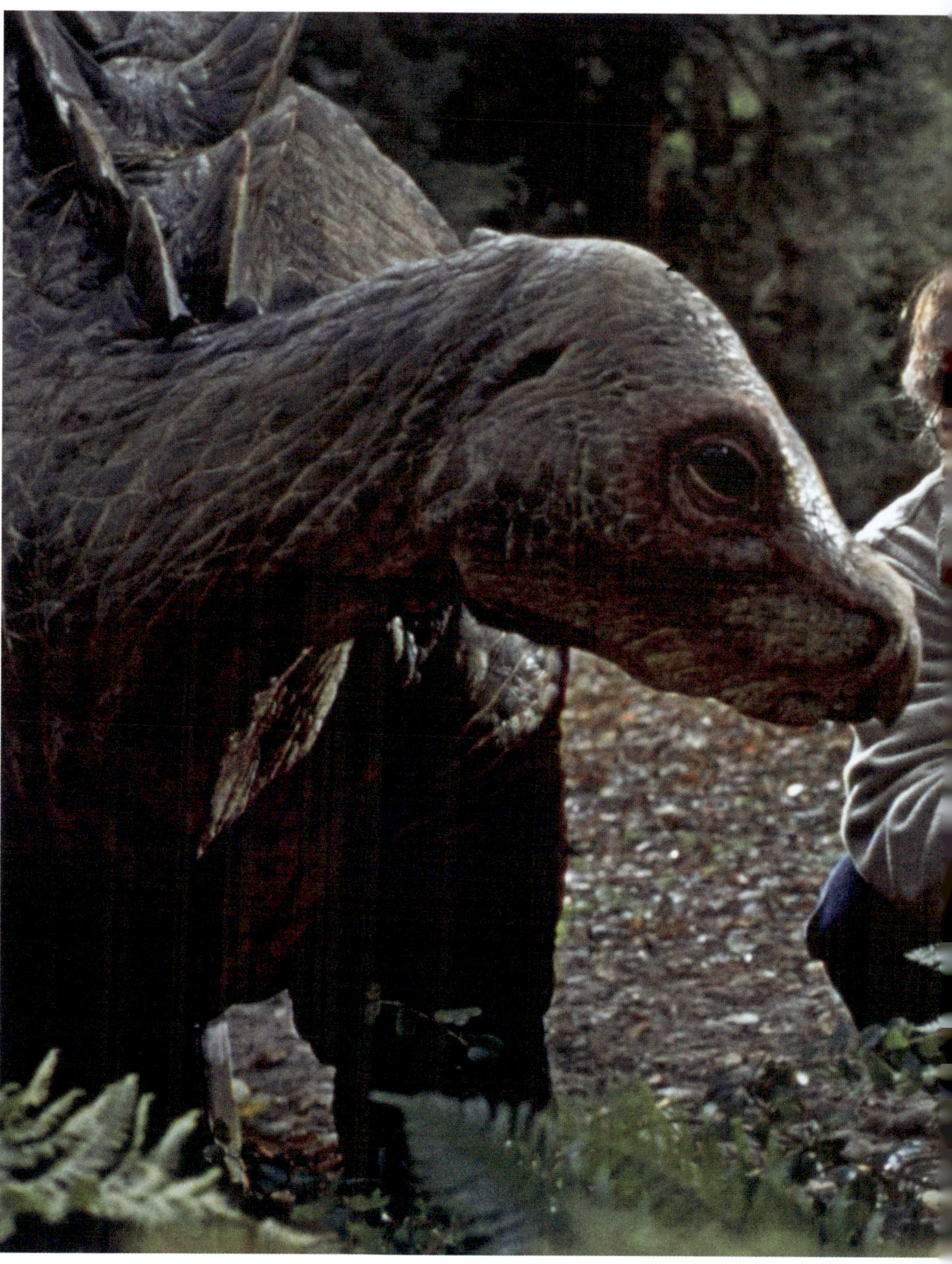

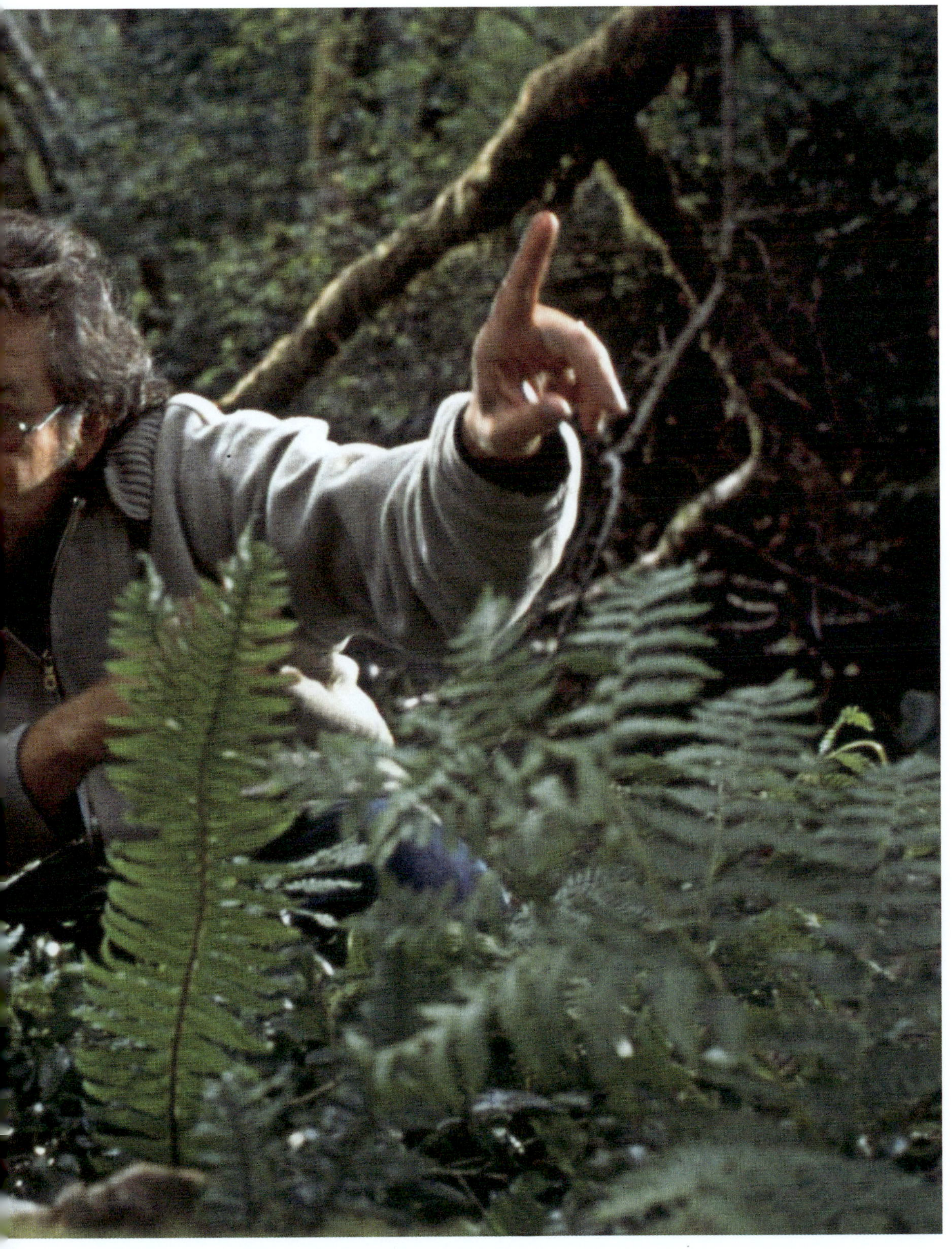

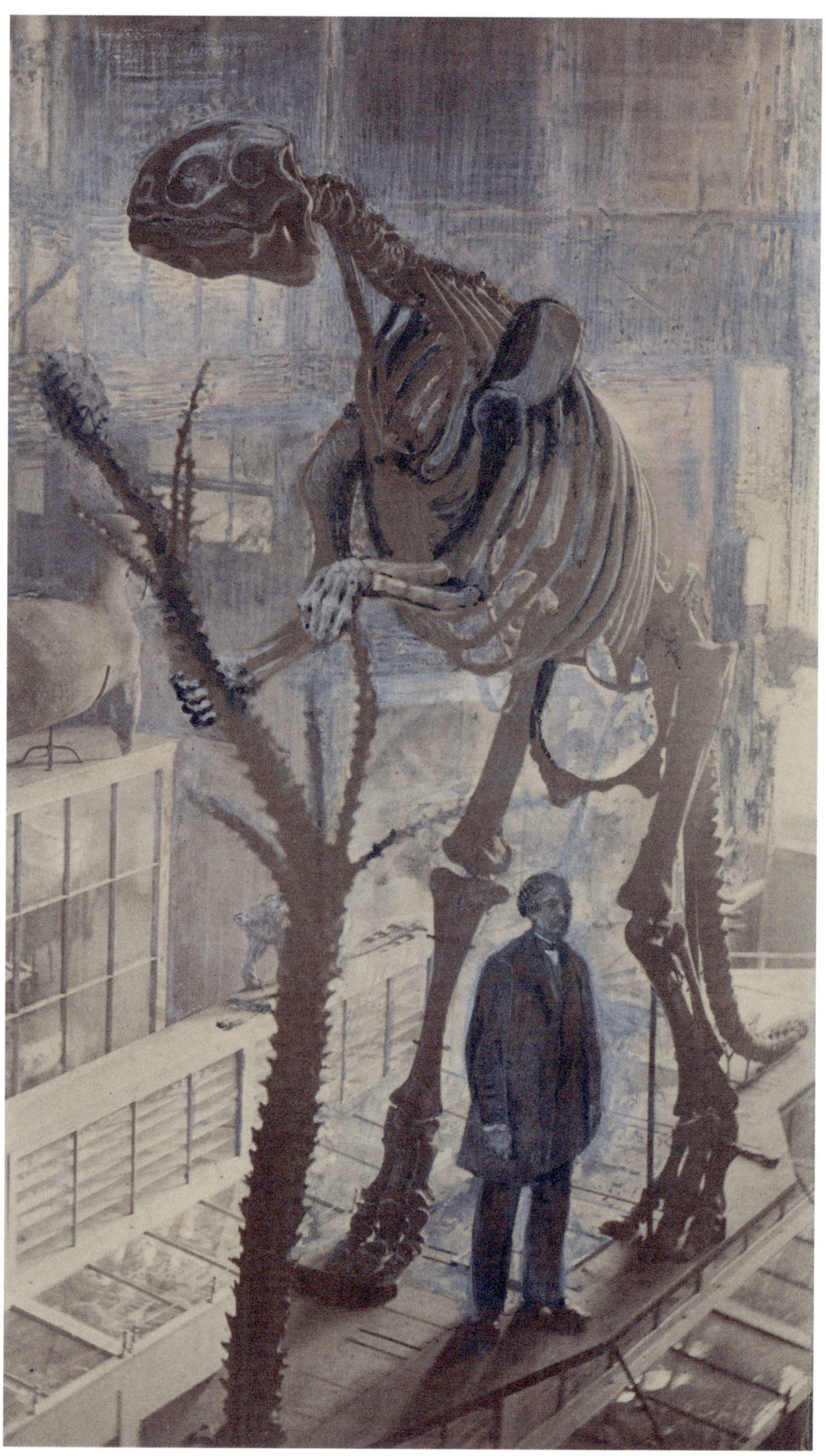

WANTED: DINOSAUR DNA OR FOSSIL BONE CELLS

In fact, what we'd actually need to recreate the dinosaur genome is a dinosaur perfectly preserved in amber, with intact DNA. In Myanmar in 2016, the plumes and tail of a small, feathered dinosaur discovered in amber were almost a 100 million years old. Even traces of muscle and skin were discernible under the microscope. So far, however, these finds have not provided suitable genetic material.

We can look for ancient DNA in other places, too. In 1966, Roman Pawlicki's team at the Jagiellonian University, Kraków, found fossilised bone cells, blood vessels and collagen fibres in dinosaur bones. Collagen is the most common protein in connective tissues such as skin, teeth and bone. A few decades later, his team provided the first evidence of 70-million-year-old DNA.

Mary Schweitzer and her team at North Carolina State University in Raleigh have been creating a furore in molecular palaeontology since 2005. They mainly use cell biological staining techniques to identify blood vessels and bone cells in bones tens of millions of years old from all kinds of extinct vertebrates. Such fossilised bone cells are more common than you might think. They often contain proteins and DNA. Unfortunately, the DNA is so corrupted that it yields no useful data, making it virtually impossible to extract a usable DNA sequence from which you could code the animal. It's like attempting to reconstruct a novel from a few words. But the proteins do appear to be reasonably stable over time. They have a measurable structure that can be compared to that of recent organisms. The cells and their contents are best preserved in oxygen-rich environments, such as sandstone. Oxygen and iron in the bones cause chemical reactions that make the proteins more resistant to water and bacteria, preserving them, intact, for hundreds of millions of years. This is why molecular palaeontology is so fascinating: on the one hand, it helps us better understand exactly how bones fossilise and, on the other, it opens up a world of possibilities for evolutionary studies. But scientists are unlikely to reconstruct a dinosaur from fossil DNA anytime soon.

———

<

Benjamin Waterhouse Hawkins mounted the first dinosaur found on American soil,
the *Hadrosaurus foulkii*, in 1868.

HOW TO BUILD A DINOSAUR

Could it be that we're taking things a step too far? Why do scientists insist on resurrecting dinosaurs when they might still be here? And we don't mean a scenario like in Arthur Conan Doyle's novel *The Lost World* (1912) or *The Land That Time Forgot* (1918) by Edgar Rice Burroughs – searching for an unspoiled place on Earth where prehistoric animals still roam wild. Maybe we can make do with studying birds? After all, they evolved from a group of small carnivorous dinosaurs, related to the well-known *Velociraptor*. As they evolved, they lost some of their ancestral traits: teeth in the beak, a long tail or 'fingers' ending in claws. Yet, remarkably, when a bird embryo matures in the egg, these features appear briefly. For example, during the first stage of development, a bird embryo still has separate 'fingers' tipped with claws, but as it develops, these bones fuse together to form a hand with important attachment points for flight feathers. In the tail, the bones also merge to form a small stump, the pygostyle. In modern birds, this tail has almost as many vertebrae as the long bony tail of *Archaeopteryx*. You can intervene in these developmental processes if you know the underlying genetic signals. During development, the activation of certain genes causes these deformities. If you can suppress the expression of these genes, a bird could develop with claws on its hands, a long tail and other features typical of theropod dinosaurs. The same technique could also be used to manipulate tooth development.

>

Is *Archaeopteryx* the prehistoric cross between a reptile and a bird?

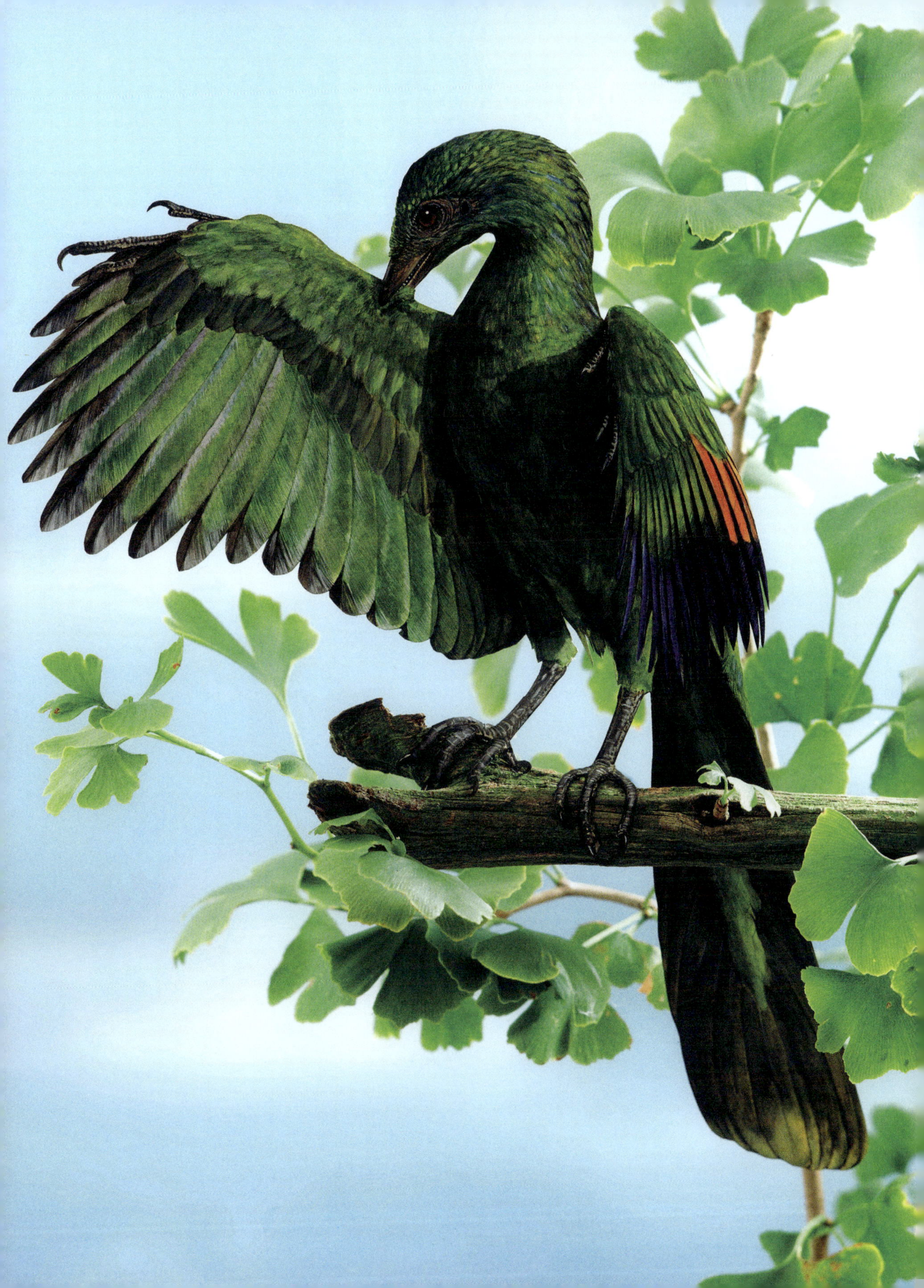

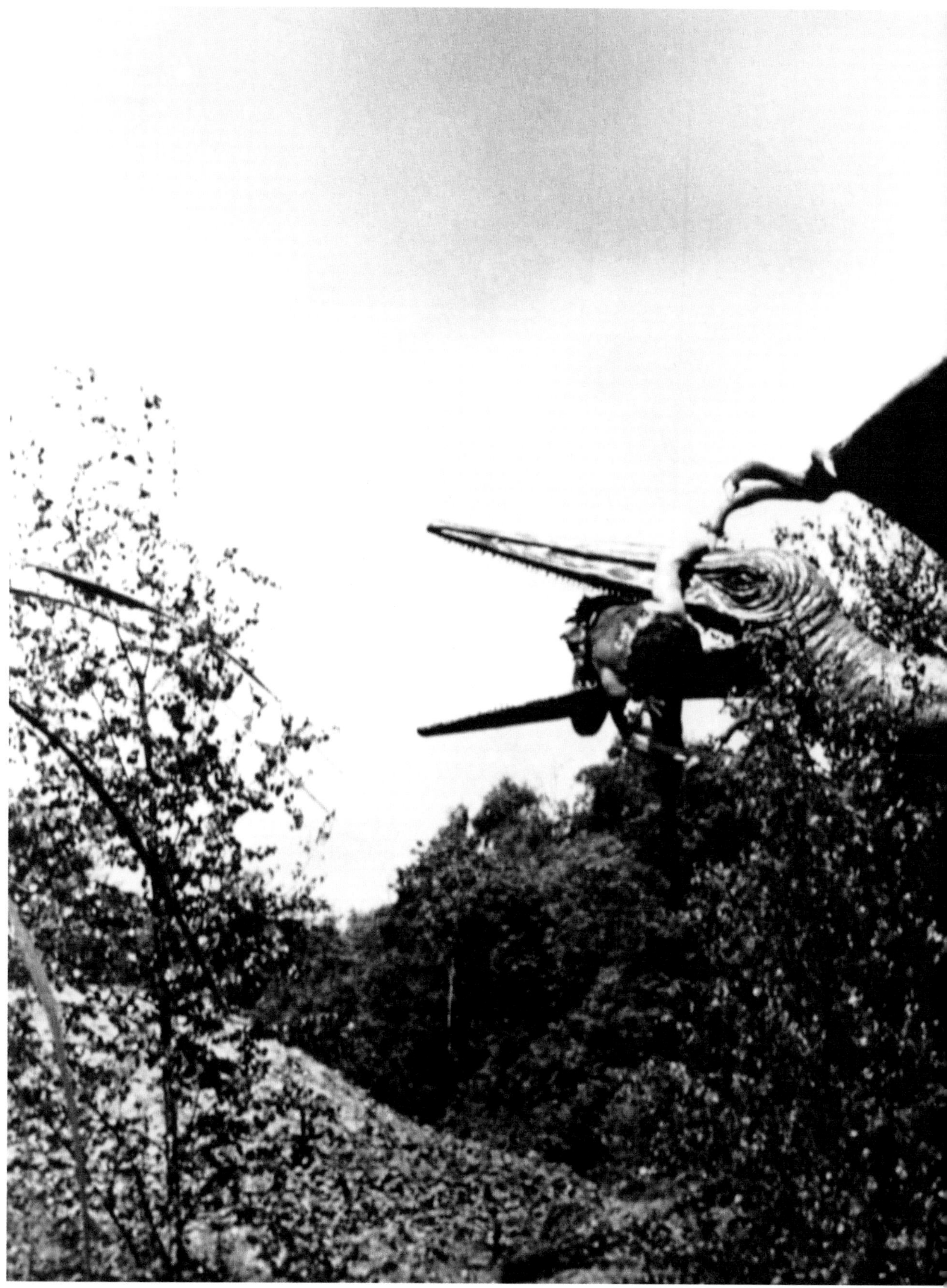

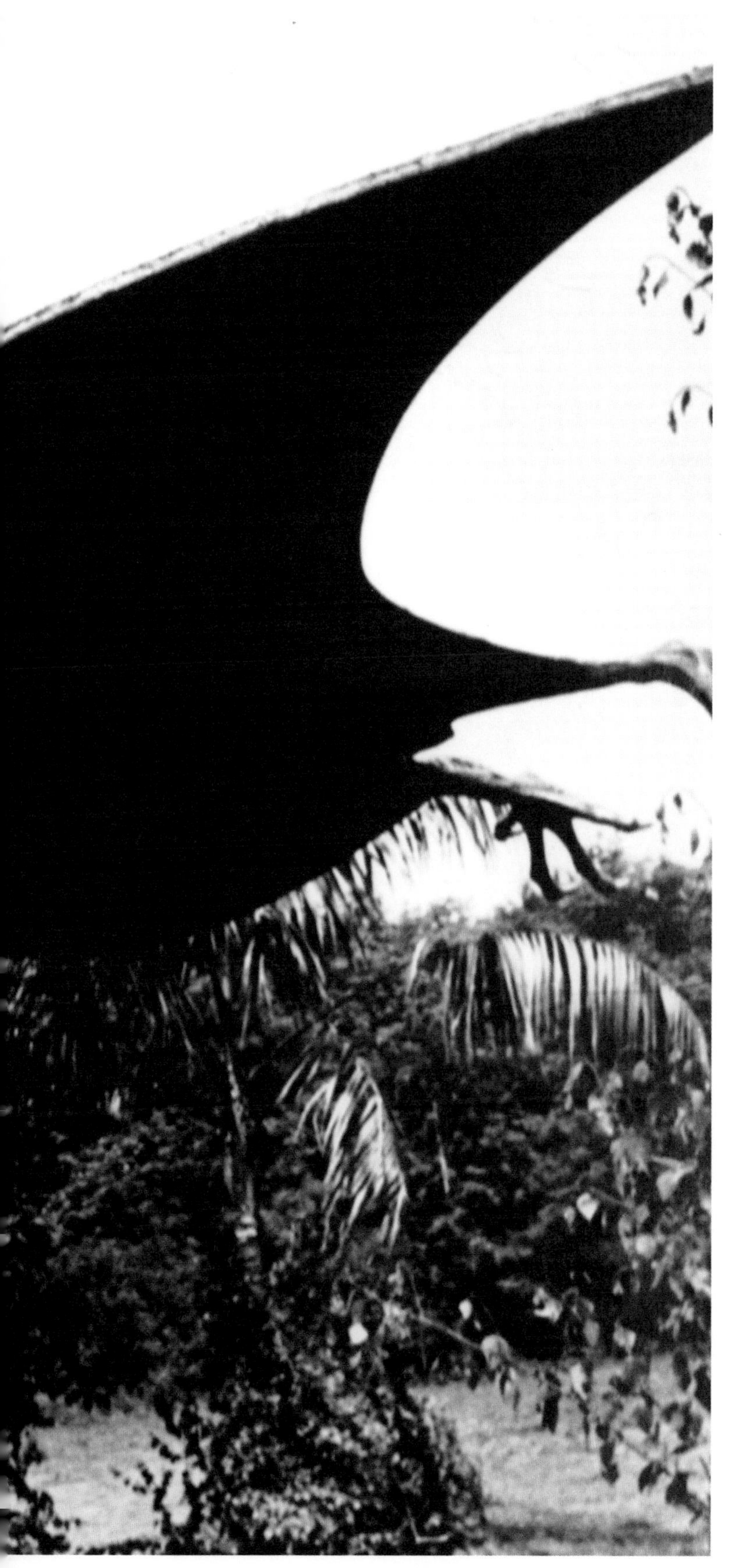

Film still from *The Land That Time Forgot* (1975):
the dinosaur film that became a cult classic.

THE DINO-CHICKEN OR CHICKENOSAURUS

In 2010, American palaeontologist Jack Horner wrote a fascinating book about intervening in chicken embryo development: *How to Build a Dinosaur.* In the book, he sets out the process that he intends to use to engineer a 'dino-chicken' or 'chickenosaurus'. Horner chose the chicken because it is a bird that is commonly used as a scientific research model. Meanwhile, most of the steps in his plan for the 'dino-chicken' have been completed. His experiment fails – as is often the case – when it comes to financing, as well as with respect to the tail and the tooth enamel. Birds no longer seem to possess the latent genes for a long tail or tooth enamel. This data was completely wiped out during the evolutionary process. To engineer a long-tailed dino-chicken, the researchers will need to change direction. They can implant genes from other long-tailed reptilians descended from dinosaurs, such as crocodiles. But by doing so, surely they'd invalidate Horner's premise? Because Horner's actual goal is to retro-engineer a chicken to demonstrate the principles of evolution to the general public. And this didactic aspect of the project is vital because, in the United States, many people are opposed to the teaching of evolution in schools.

What, you might ask, is the added value of Horner's dino-chicken story? As spectacular as his idea is, by cutting and pasting genes, he seems to be missing the point. If Horner's retro-engineered dino-chicken reproduced, it would still pass on ordinary chicken genes to the next generation, because the changes are not permanently encrypted in its DNA. And what about the ethical issues raised by his research? Fortunately, an ethics committee always rigorously screens studies and experiments of this kind. The development of a chicken embryo that has been modified to express ancestral characteristics is always halted at an early stage – mostly to prevent unnecessary animal suffering but, of course, also to prevent these genetic samples from being released into the wild. So even if the idea becomes technically viable, you are unlikely to see a dino-chicken running about in the near future.

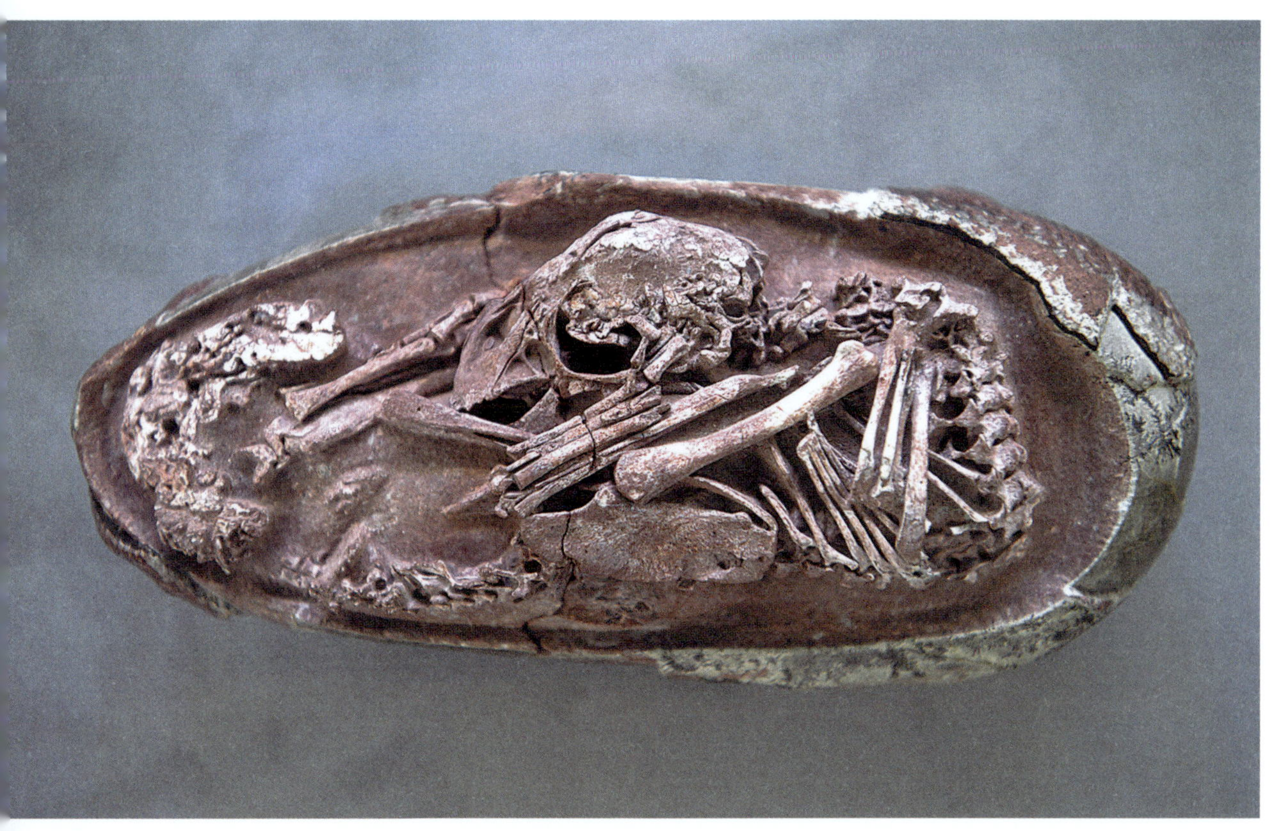

Dinosaurs laid eggs with very thin shells.

WHAT DO DINOSAUR EGGS TEACH US?

Dinosaur eggs containing embryos reveal a host of new insights about dinosaur reproduction and growth. The first eggs with *Maiasaura* dinosaurs foetuses were described by Bob Makela and Jack Horner in the late 1970s. Meanwhile, numerous eggs containing embryos have been discovered, including specimens that could be from a therizinosaur, *Oviraptor*, the prosauropods *Massospondylus, Mussaurus* and *Lufengosaurus*, titanosaurs from Argentina and *Protoceratops*. Dinosaur eggs and eggshells turn up quite frequently, but can only be attributed to a species if bone material can be sourced from the embryo. Palaeontologist Koen Stein – co-author of this book – and his team examined eggshell fragments of the prosauropods *Massospondylus, Mussaurus* and *Lufengosaurus* under the microscope. Using various chemical methods, the team discovered that the first dinosaurs laid eggs with very thin shells, no thicker than a human hair. In the course of their evolution, several large groups developed thicker eggshells, but all had different structures and compositions. The evolution of thicker eggshells would have occurred no earlier than the Jurassic, possibly linked to an increase in atmospheric oxygen.

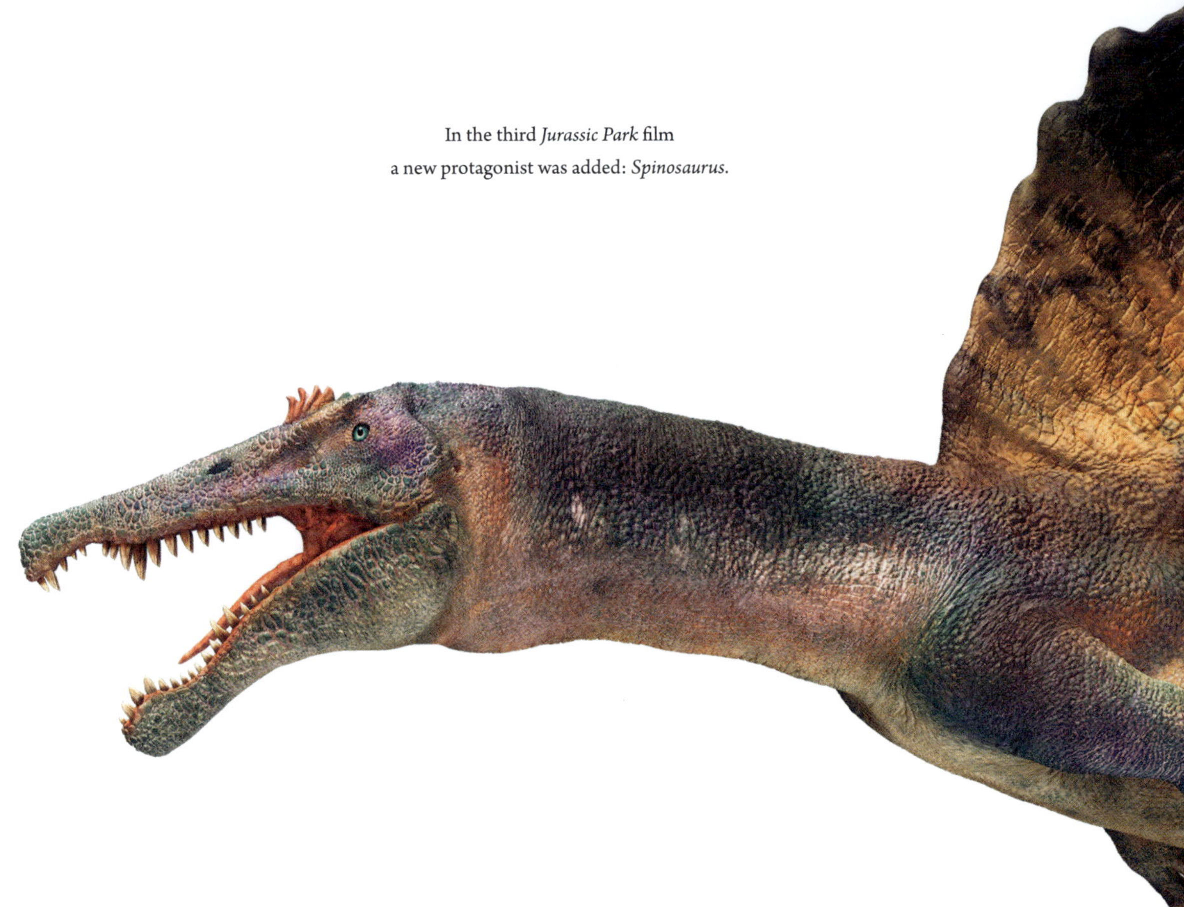

SPINOSAURUS IN *JURASSIC PARK*

The dinosaur revolution was undoubtedly fuelled by *Jurassic Park*. Just as Indiana Jones made public opinion about archaeology less stuffy, *Jurassic Park* made palaeontology more adventurous. Despite the many scientific inaccuracies (such as naked raptors), the 1990 book and the first 1993 film were enough to keep an entire generation enthralled by dinosaur research. *Tyrannosaurus rex* and *Velociraptor* were the protagonists of the first two films. Only in the third sequel from 2001 was a new protagonist added: *Spinosaurus*. This was inspired, among other things, by new discoveries in the 1990s. Also, after 2000, there were some sensational new *Spinosaurus* discoveries in Tunisia, Algeria and Morocco. In 2014, a team led by Nizar Ibrahim even launched a new reference type to replace the holotype specimen from Egypt, which had been destroyed during the Second World War. The new skeleton provided much additional information about the physique and size of the *Spinosaurus*. But it resulted in fierce reactions from palaeontologists. The skeleton came from Morocco, which is more than 3,000km away from where the original skeleton was found. Doubts remain as to whether the new skeleton represents a single individual. It has strange proportions, with very short hind legs, which would indicate a swimming lifestyle. Scientists still do not know whether the animal swam or waded through life, like a monstrous heron.

———

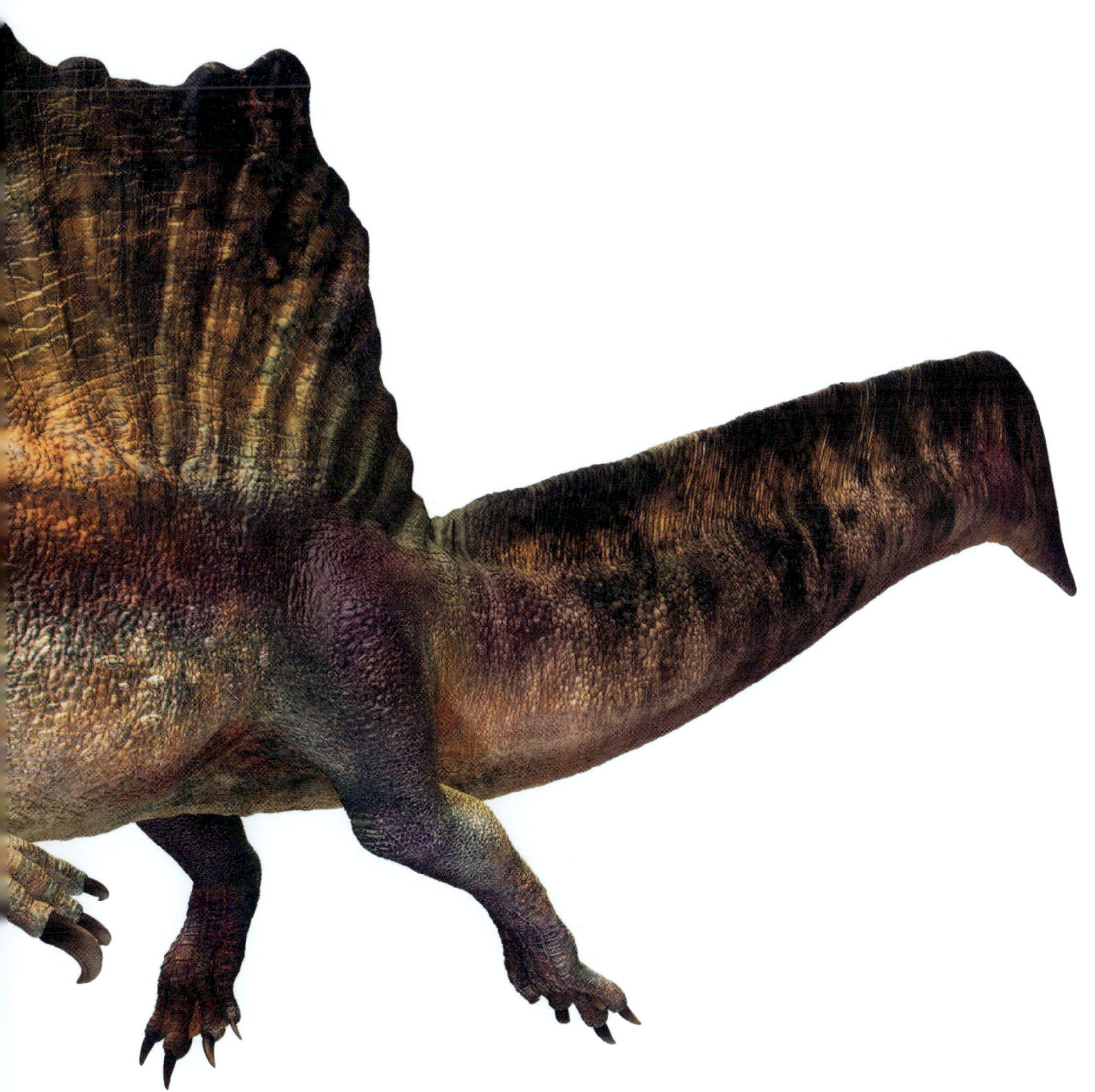

THE BEST-PRESERVED DINOSAUR

Borealopelta markmitchelli is the best-preserved dinosaur to date. The find was described in 2017, but actually the fossil was found back in 2011. It was preserved in an incredibly hard marine limestone. This made preparation very time-consuming, but ensured excellent preservation. The fossil is a fossilised mummy of a nodosaurid ankylosaur. In human language: an ankylosaur, but without the tail club. The entire dorsal armour, consisting of ossified skin plates (osteoderms) is still completely connected, and on these plates there are still horn layers and possibly even remnants of the animal's colour. No doubt we will hear much more about this animal in the future.

A NEW SPECIES EVERY WEEK

Although palaeontologists have been working on the taxonomic classification of species since the discovery of dinosaurs in the mid-19th century, it is by no means the case that all dinosaur species have been discovered in the meantime. There are periods when new dinosaur species are published every month or even every week. This does not make it easy for palaeontologists to keep up to date. The book *The Great Dinosaur Discoveries* written by the British palaeontologist Darren Naish in 2009 is illustrative. In it, he discussed some of the most important discoveries up to 2009. In 2013, he wrote in a blog post that his chapter 'The 21st Century Dinosaur Revolution' was already very out of date.

RESEARCH ON DINOSAUR GIGANTISM

Palaeontology is still rapidly evolving, especially now that there is more cross-fertilisation with other disciplines, such as genetic research. Thanks to new technologies like microtomography and synchrotron radiation, we have gained a lot of new insights into the biology and evolution of these fascinating animals in recent years. The latest research on the size of sauropods, such as *Diplodocus, Camarasaurus* or *Brachiosaurus*, also proves that collaborations between scientists from different disciplines pays off. At the University of Bonn, from 2004 to 2015, an interdisciplinary research group consisting of some 20 palaeontologists, zoologists, food scientists, geochemists and materials experts looked at how, and why, these creatures got so big. How had their huge bodies functioned? The researchers got to work, trying to discover how much these giant dinosaurs ate in a day. If sauropods were the giant equivalent of elephants, they would have consumed and excreted ten times as much: 2.7t and 700kg per day, respectively, the zoologists calculated. By tackling the problem of gigantism from multiple scientific angles, the team headed by Martin Sander unravelled the characteristics that allowed sauropods to reach such a momentous size. The secret proved to be a combination of ancestral traits (such as egg-laying) and newly evolved biological features (such as a complex lung with a bellows system of air sacs like modern birds) that were advantageous to the common ancestor of all sauropods. It led to an accelerated increase in size, an evolutionary step supported by the abundant food available on the supercontinent Pangaea and, after the Jurassic, also on Laurasia and Gondwana. The principles identified by the research group were also confirmed by the dwarf sauropods *Europasaurus* from the Jurassic of Germany and *Magyarosaurus* from the Cretaceous of Romania. These creatures lived on islands and were no larger than a horse, whereas their relatives on the mainland are among the largest animals that ever walked the Earth. Island-dwelling creatures have limited food resources and would have evolved into miniaturised forms within about 10,000 years.

BIG DATA IN DINOSAUR RESEARCH

Not only interdisciplinary research but big data is seeping into modern palaeontology. Using the latest computer technology, many researchers are now able to study dinosaurs' evolutionary patterns. One of the resources they use is the Paleobiology Database (www.paleobiodb.org), which, since 1998, has brought together a growing number of digitised collections. The platform is open source, and can be used by anyone wishing to carry out (data) research. Almost 400 scientists from more than 130 different research institutes in 24 countries have added information to the database for 25 years. Needless to say, it is now gigantic. The database was actually created to document the evolution of biodiversity through geological time, to facilitate researching topics such as the extinction and spread of new species.

The Paleobiology Database allows you to visualise certain species over certain time periods. You can find them in collections shown on a contemporary world map. Or in their habitat on a world map from the past. Researchers can also distil graphs from it about how the biodiversity of certain groups has changed over time. Meanwhile, a countless number of scientific publications use information from the Paleobiology Database. It continues to be a highly reliable source for people looking for primary literature on certain dinosaur (or other) species.

———

>

Drawing by Gustave Lavelette showing the exact position of
an *Iguanadon* skeleton in the Bernissart mine.

. Reg. des types : E.F.R. 56 . - Anc. reg. os. fos. 1561

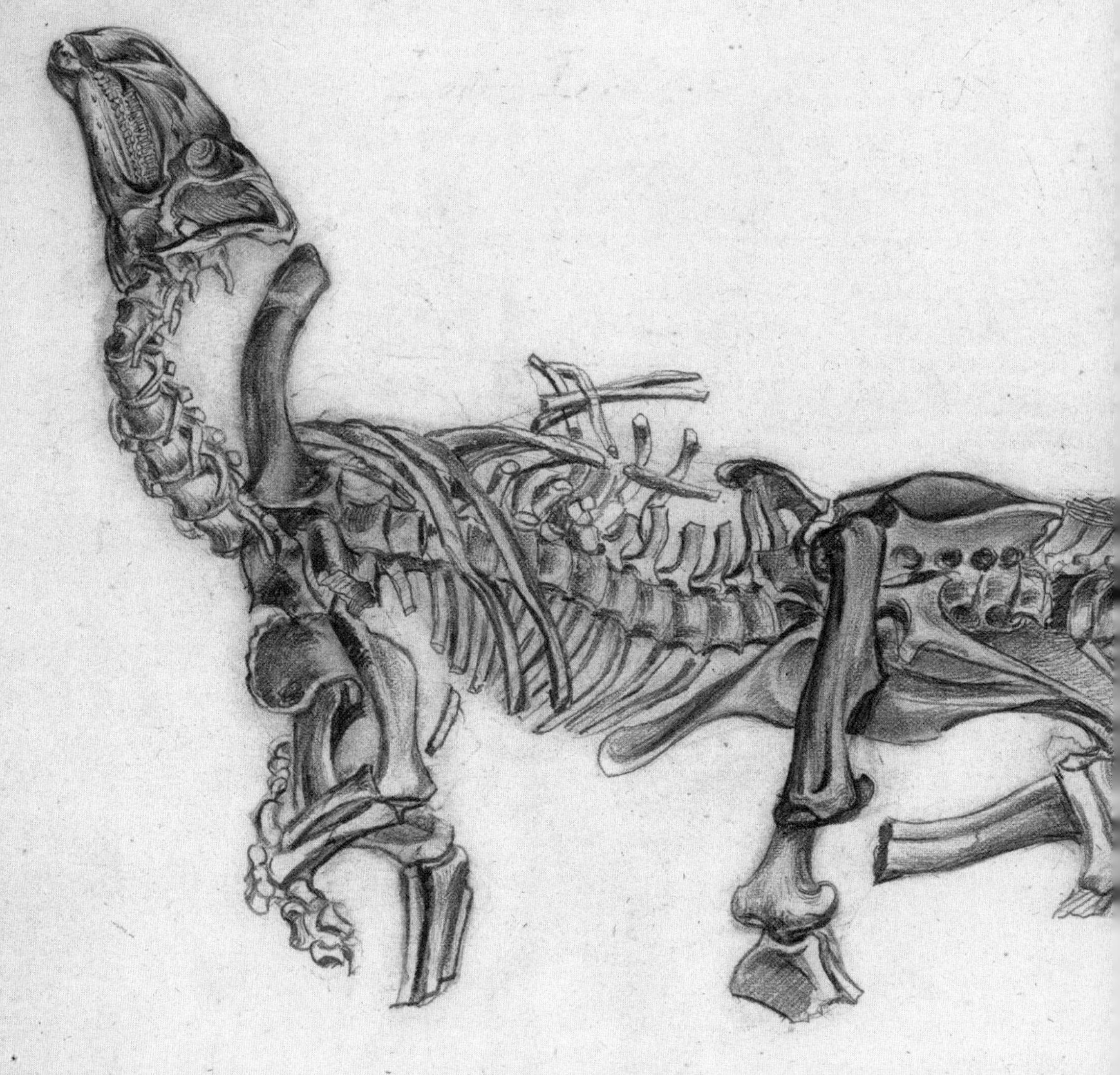

Planc. 7. Coupe A.B.

Lettre. L.

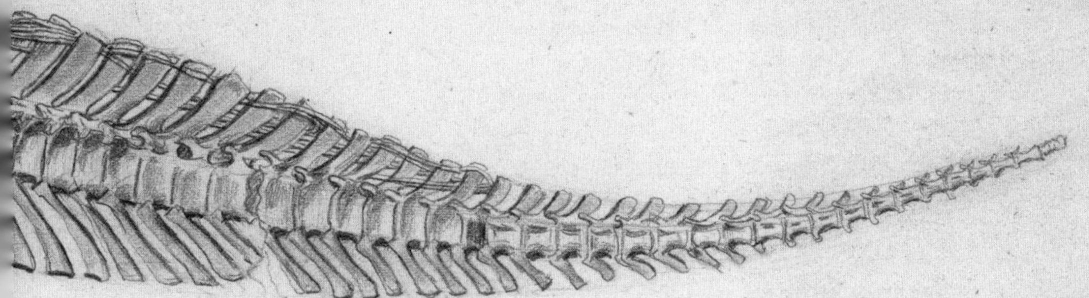

Gve. Lavalette

FIRST FEATHERS, THEN FLIGHT

The first dinosaur fossils with feather-like structures were discovered in the 1990s (not counting the 19th-century discovery of *Archaeopteryx*). Other finds soon followed. Some, but not all, of these fossils were directly related to birds. In the 21st century, our knowledge of feathered dinosaurs deepened, with the excavation of theropods such as *Sciurimimus* in Germany and *Yutyrannus* in China, as well as *Kulindadromeus* in Russia, which is not a theropod, but an ornithischian. These finds reveal that dinosaurs developed downy coverings and plumage long before they could fly. Feathers may have their genetic origin in a common dinosaur ancestor. There is plenty of evidence that dinosaurs were warm-blooded. A feathery covering was probably a survival mechanism, providing thermal insulation to maintain body temperature. At a later evolutionary stage, feathers also proved to have other uses, such as flight.

NEW INSIGHTS INTO THE EXTINCTION OF THE DINOSAURS

New evidence about the abrupt end of the dinosaur era is still coming to light. We have known since 1980 that a massive meteorite or asteroid slammed into Earth, wiping out the dinosaurs. In 1991, the impact crater was discovered on the Yucatan Peninsula. Roughly 180km in diameter, the crater formed when an object 10–15km wide struck the Earth, causing chaos and disrupting the biosphere. At least 75 per cent of all species on Earth became extinct. Many believe that the only creatures able to survive were those weighing less than 20–25kg. However, until 2013, it was thought that the mass extinction occurred 65 million years ago. Paul Renne and his team have pinned the date down more precisely, establishing that the event occurred 66 million years ago. Launched in 2016, the Chicxulub Crater Drilling Project, which involves drilling deep into the crater, should yield greater insights on the scale of the impact, the meteorite, and its effects on Earth's environment.

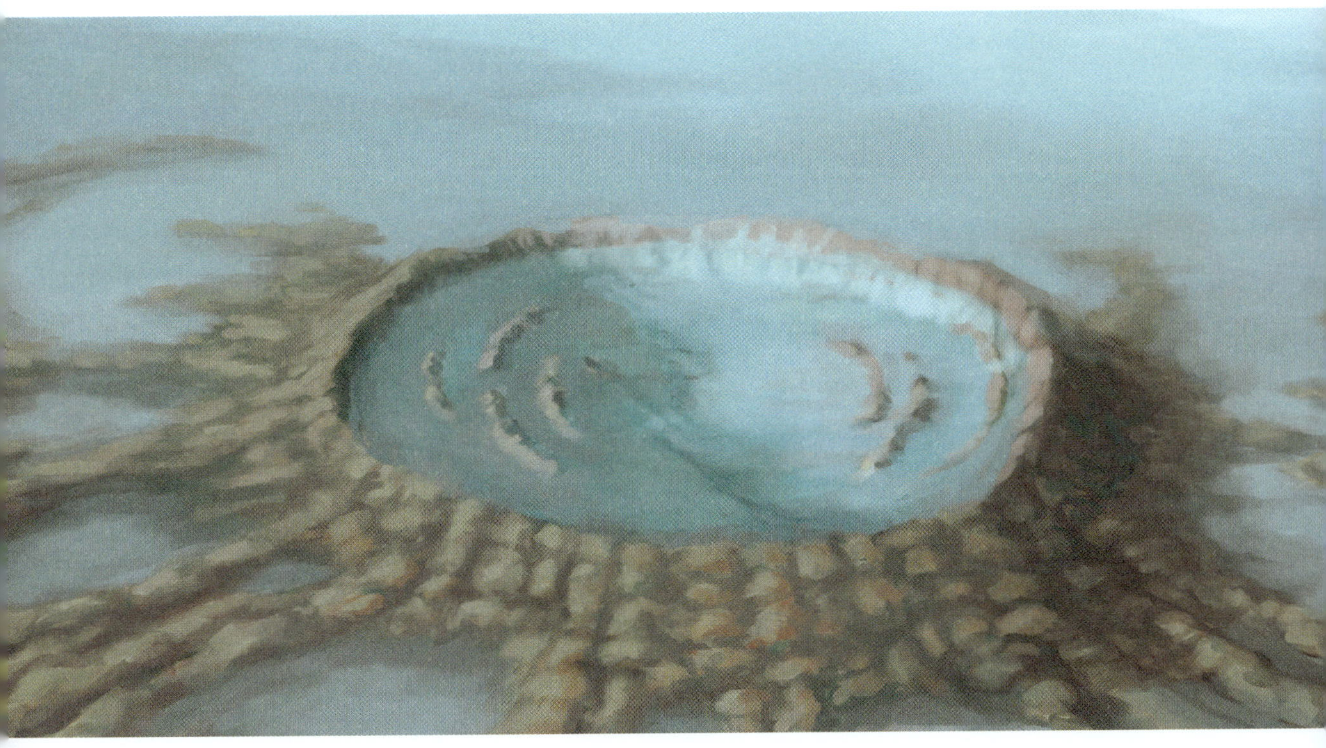

A comet 10–15 km in diameter struck the Yucatan Peninsula,
destroying 75 per cent of life on Earth.

A great deal of new data on the dinosaur extinction has been unearthed in
recent years. Finds from the far east of Russia, including *Amurosaurus*, *Olorotitan* and other hadrosaurs, described by teams led by Belgian palaeontologist Pascal Godefroit between 2004 and 2012, clearly show that the mass
extinction at the end of the Cretaceous was abrupt, not gradual, as some had
claimed. *Olorotitan* and *Amurosaurus* lived at the end of the Cretaceous and
belong to a large group of hadrosaurs with a cranial crest, like *Parasaurolo-
phus*. This group virtually disappeared from North America long before the
end of the Cretaceous Period. We now know that this gradual extinction in
North America was a local phenomenon, probably caused by the uplift of the
Rocky Mountains and the disappearance of the Western Interior Seaway.

———

>

Fossil hunters Clayton and Luke Phipps in the Hell Creek Formation, Montana,
where scientifically important Cretaceous dinosaur remains
have often been found.

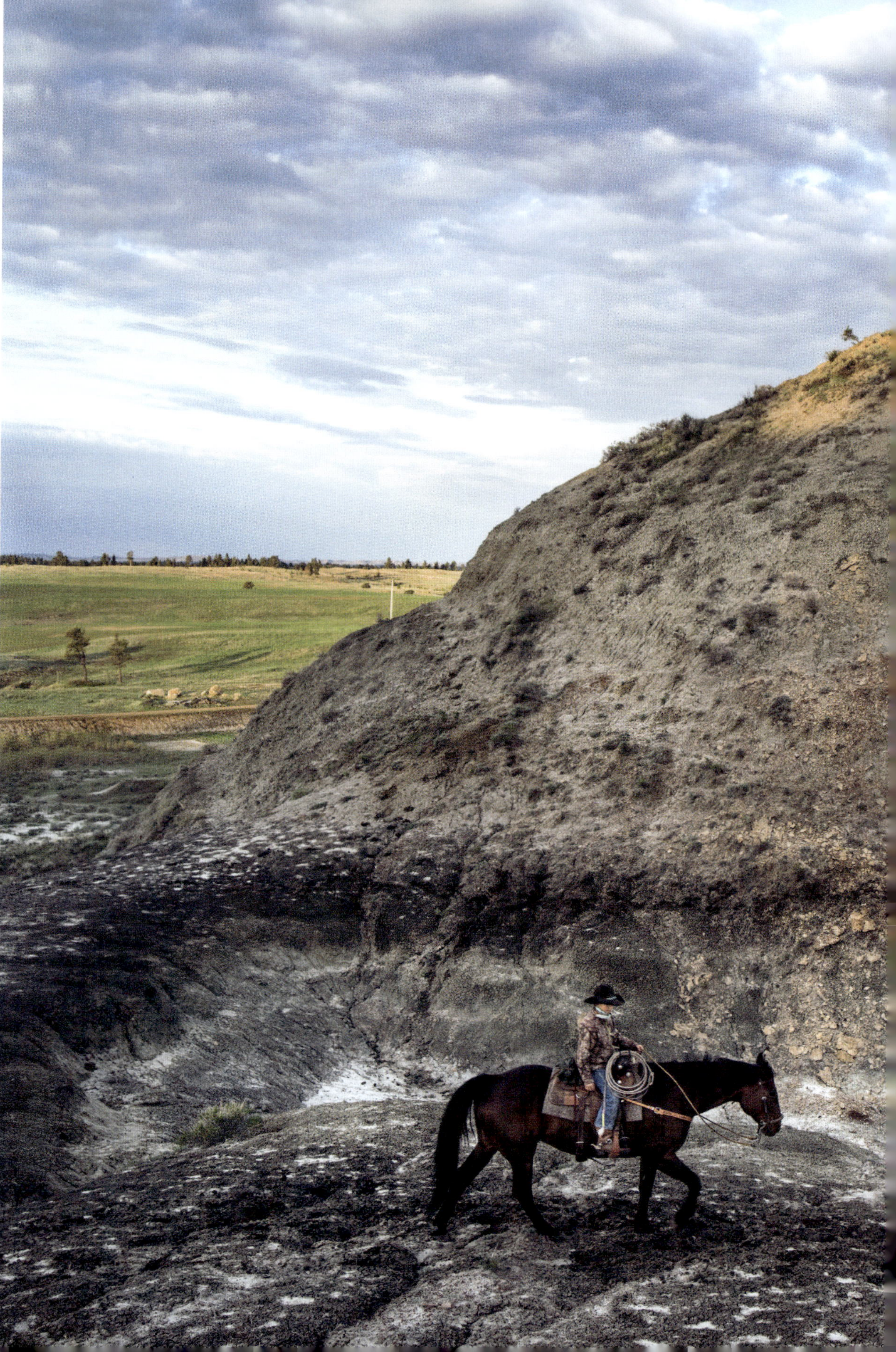

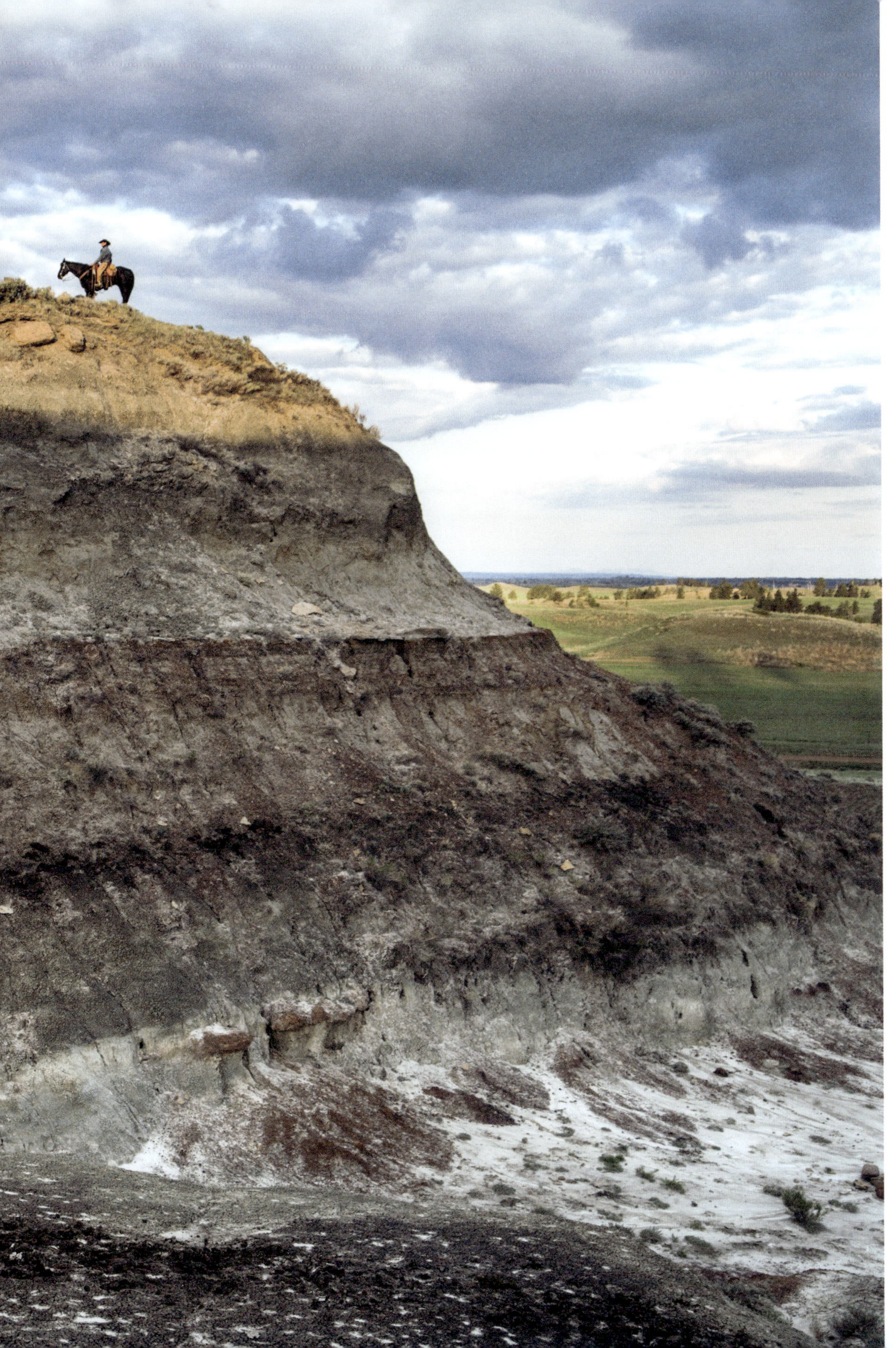

THE POMPEII FOR PALAEONTOLOGISTS

The Pompeii of dinosaurs may well be in North Dakota. When the Tanis fossil site was discovered in 2019, the news caused a sensation. The site is unique in that it records events from the first minutes to several hours after the fatal impact. The site, described by a team led by geologist Robert DePalma, is part of the Hell Creek Formation, where the transition from the Cretaceous to the Paleocene is clearly visible. A jumble of plants, land animals, freshwater and saltwater fish, as well as bones and even feathers from all kinds of dinosaurs, are piled up at random. Among these remains are also tiny spherules of rock (microtektites) hurled into the atmosphere by the impact of the comet or meteorite. Traces also turn up in fossilised fish, which choked to death when impact grit congested their gills. The location is too remote from the crater (around 3,000km) to have been struck so swiftly by mega-tsunamis, leading to theories that the vegetation and creatures were flung together by 'seiches', or standing waves, in the Western Interior Seaway, the sea that bisected America from Canada to the Gulf of Mexico. Seiches are thought to have been caused by seismic activity in the Earth's crust. A seiche is a standing wave in an enclosed body of water. It can best be compared to a bucket of water shaken rhythmically until the water splashes out. In 2022, a team led by the Dutch palaeontologist Melanie During was even able to determine the exact season of the catastrophe. By studying the growth rings in the gill covers of fish from the site, they determined that all of the creatures had died at the start of the growing season: spring. The Tanis site has only just been described and is likely to be a subject of study and research for decades to come, presenting ample opportunities for young palaeontologists to make new discoveries.

THINLY SLICED DINOSAUR BONES

For many palaeontologists, discovering a new species is the pinnacle of their career. But sometimes well-known fossil species also provide new insights if you look at them differently. As science evolves, palaeontologists can use new methods, techniques and technologies, often from other disciplines. The Belgian–French palaeontologist Armand de Ricqlès started examining bone tissues from dinosaurs and other extinct reptiles under the microscope in the 1960s. He used his histology expertise to learn more about their physiology. Taking small samples of dinosaur bones, mainly from the collections of the Musée National d'Histoire Naturelle in Paris, he ground the samples until they were translucent so he could study the fossilised tissues. He carried out several systematic studies, finding evidence to prove that dinosaurs grew far too quickly to be cold-blooded. His work is one of the foundations of modern palaeobiology, a branch of palaeontology that is devoted to better understanding the physiology, growth and biological functions of extinct organisms.

With so many fossils to (re)discover in the existing collections of the world's natural history museums, is there any point in digging for new ones? Absolutely, because in order to deepen our knowledge of the changing biodiversity and evolution of life on Earth, we need to go on digging for fossils. But our existing collections must also be properly cared for, as a potential source of new insights and knowledge. Collections form the basis of palaeontological knowledge. The application of new techniques from other disciplines, such as chemistry and physics may also generate a wealth of new insights, and a palaeontologist must keep pace with general scientific advances. Palaeontology is alive and kicking. New studies, often produced by a multidisciplinary team, appear regularly, and doing justice to all of the newest discoveries is outside the remit of this book. Luckily, countless blogs, podcasts and YouTube channels offer regular updates.

THE NATURAL HISTORY MUSEUM OF THE FUTURE

In 1868, Benjamin Waterhouse Hawkins was the first person in the world to mount a dinosaur skeleton. The stuffed skeleton drew crowds at the Philadelphia Academy of Natural Sciences: *Hadrosaurus* really came to life (see Chapter 2). To this day, natural history museums with displays of dinosaur skeletons are a crowd magnet. And, 150 years later, modern digital techniques such as 3D printing or augmented reality give us a fun way to resurrect these prehistoric giants. In 2018, the Royal Belgian Institute of Natural Sciences developed the *Sauria* app so you can search for dinosaurs and other extinct reptiles and collect them like Pokémons on your smartphone or tablet. The Berlin Museum of Natural Sciences installed rotating binoculars in the Jurassic Period room, and if you look through them, you'll see the skeletons of the giant *Brachiosaurus* or treacherous *Allosaurus* overlaid with muscle and skin, and view these dinosaurs just as they were when they roamed in the Jurassic. Hawkins would have found it fascinating.

DIGITAL FOSSILS

Recently, there's been a surge of interest in digitising dinosaur skeletons and other important fossils. This has innumerable advantages. First, it's easier to exchange data with other researchers, who are often in other countries. Digital fossils are also a kind of back-up. Throughout history, many specimens have been lost because a collection has been mismanaged, or because of wars or natural disasters. Such losses interfere with scientific research that seeks to engage with the specimens again. A digital copy ensures that at least some basic information about the specimen is preserved and available. Bones that were deformed during the fossilisation process can be restored to their original state in digital form and, if necessary, even printed in full size. One example is the *Tyrannosaurus rex* 'Trix', excavated by a team from Leiden's Naturalis Biodiversity Center in Montana in 2013. The skeleton was missing its left hind leg. After scanning the remains, the right hind leg was digitally mirrored and printed out at full size. Bones missing from the tail were also added to the 3D model with modified reconstructions of other *Tyrannosaurus rex* specimens. A complete skeleton was printed out and shipped to Nagasaki. A second complete print was shown at an exhibition in Brussels.

The famous *Tyrannosaurus* 'Stan' also has many bastard brothers: dozens of replicas have been made for museums worldwide. The Royal Belgian Institute of Natural Sciences in Brussels also has one. There's certainly a market for casts and 3D prints of fossil bones. There are companies (such as Research Casting International) and institutes (such as the Black Hills Institute) that offer accurate casts of important skeletons. Museums sometimes buy these replicas because it is often impossible to find original skeletons at an affordable price.

>

Discovered in 2013 in Montana, 'Trix' became the first *Tyrannosaurus* skeleton
to be exhibited in the Netherlands. '*The Night Watch* of Prehistory'
can be admired at Naturalis Biodiversity Center in Leiden.

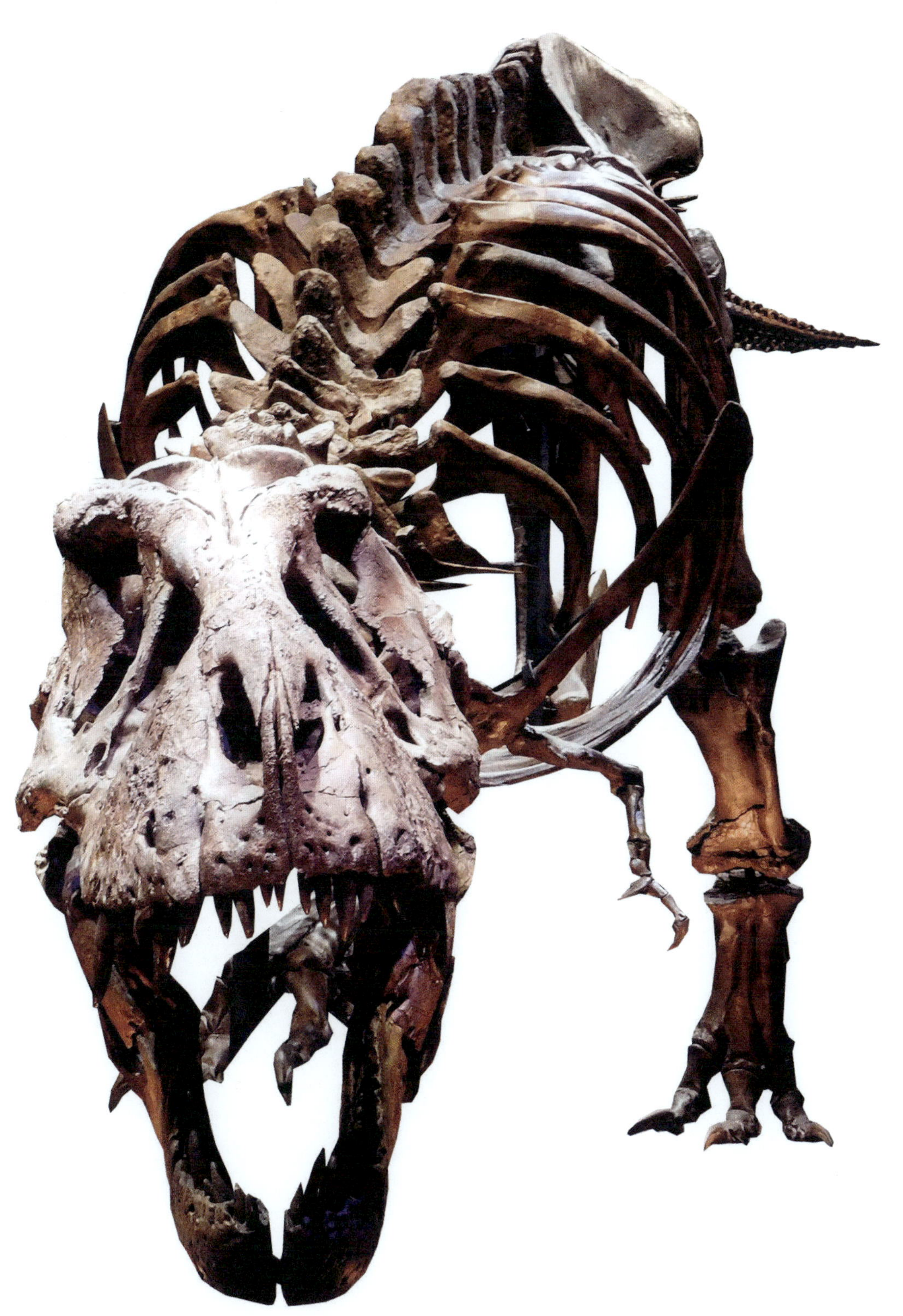

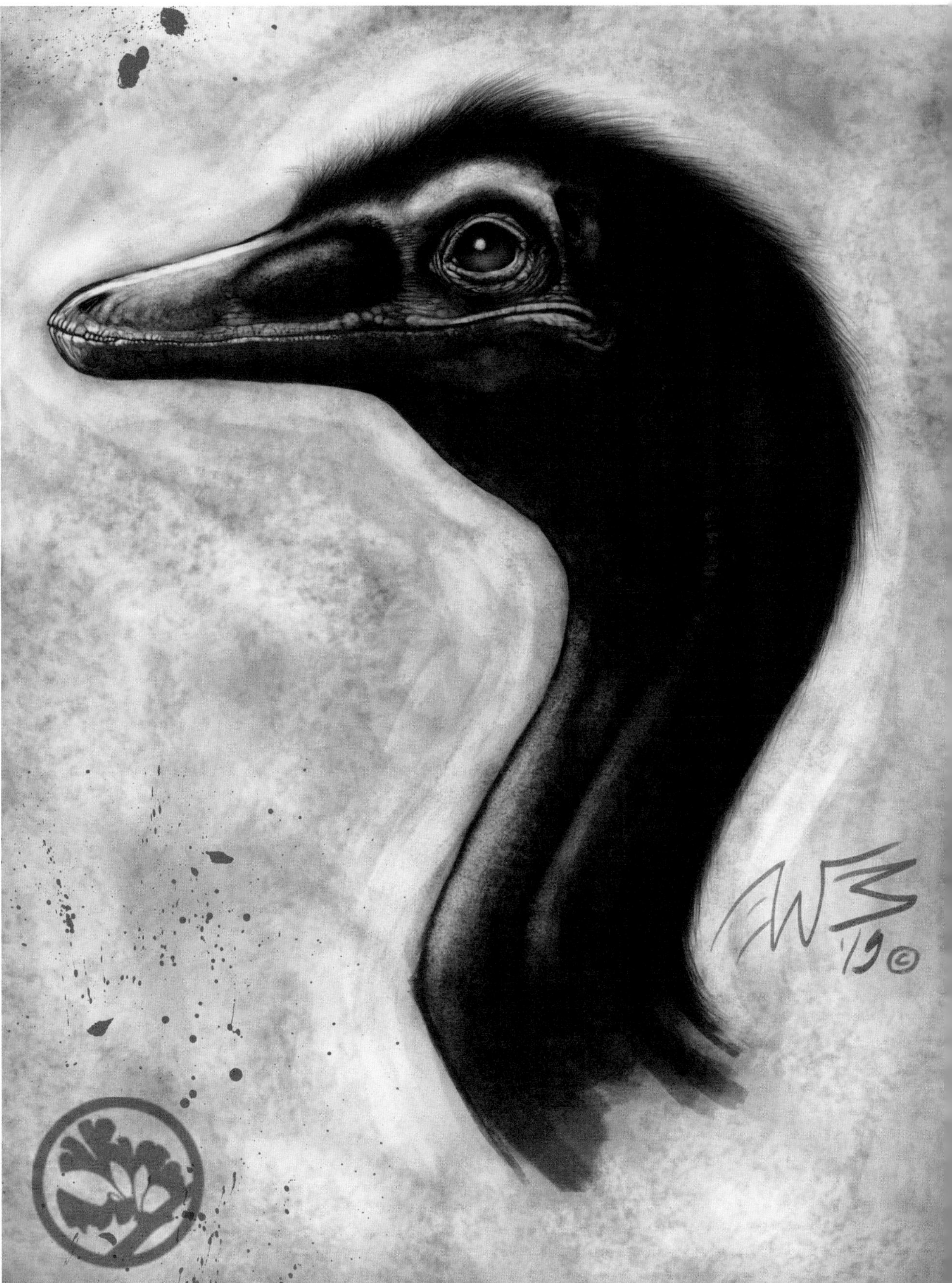

PRINTING DINOSAUR SKELETONS AT HOME

There are several ways of 3D-printing fossils. The most common are photogrammetry, laser surface scanning and (micro)tomographic (CT) scanning. Photogrammetry is the cheapest because all you need is a good camera and software. The principle is simple: you take a lot of photographs with overlaps all around the object to be digitised, including at the bottom and top. The computer software then creates a three-dimensional object based on the overlapping pixels. A laser surface scanner is a lot more expensive, but produces very accurate 3D models of fossils within a few minutes. If you want to digitise skeletons or parts of skeletons that are not too large, you can also scan them microtomographically. The advantage of that method is that you can look inside the bones because the technology works with X-rays. You can then get an idea of how the bone is built up. With the latest equipment, you can even see the cavities in which the bone cells were located, something that was previously only possible with the far more powerful radiation of a synchrotron like the one at CERN in Geneva. In any case, it opens new doors for research.

Many museums have already made their online fossil collection catalogue accessible to the public. This catalogue often contains photographs from different perspectives, but occasionally, on websites such as Sketchfab and Thingiverse, you'll come across freely available 3D models. Sometimes, through non-commercial licences, you can download models like these for free, and print them at home on a 3D printer. The museum of the future will gradually move into your living room – which might make dinosaurs collectible for everyone.

—

< >

Struthimimus (left) and *Gotcha* (following pages),
both works by Belgian paleoartist Ward Broes.

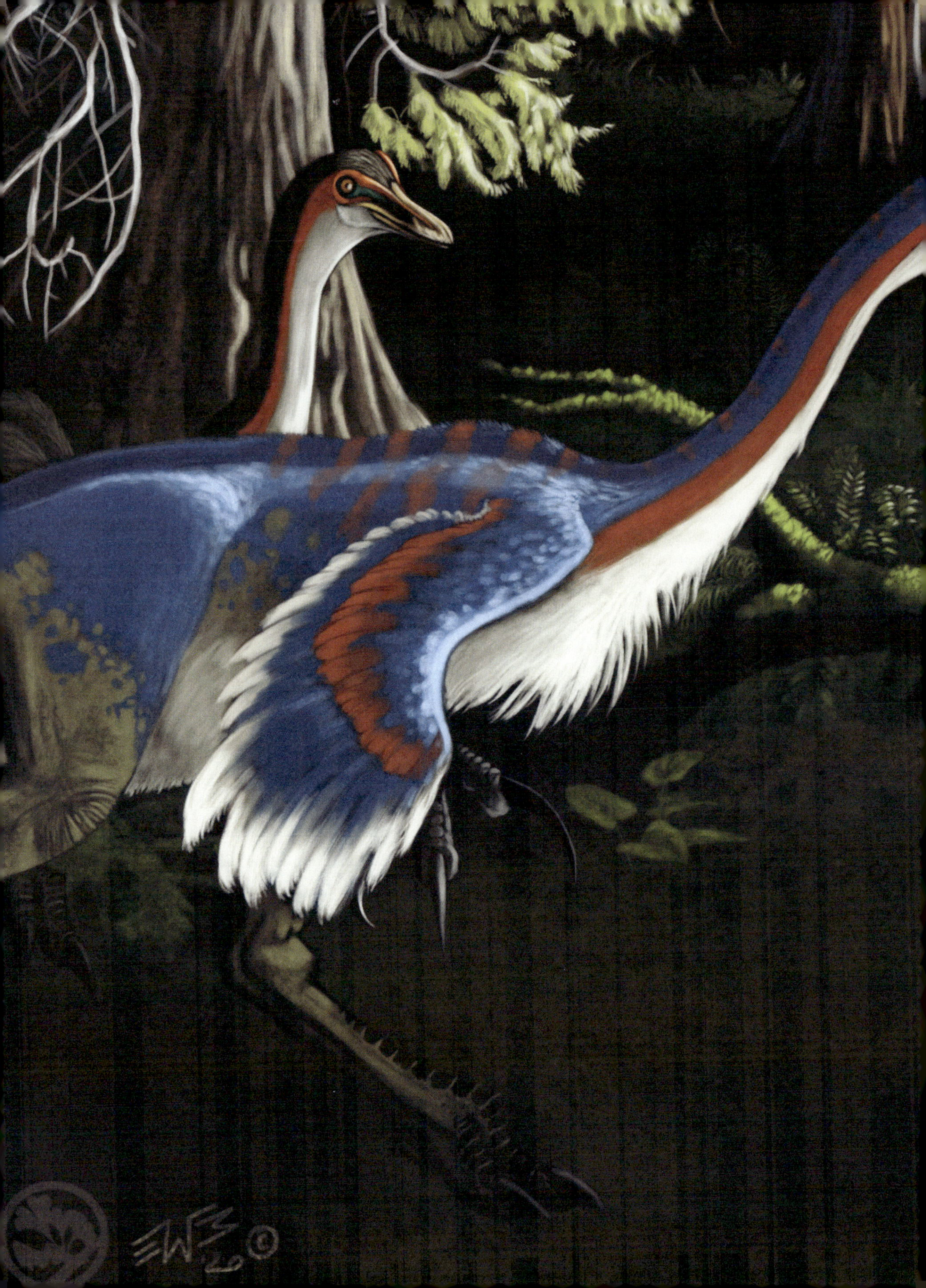

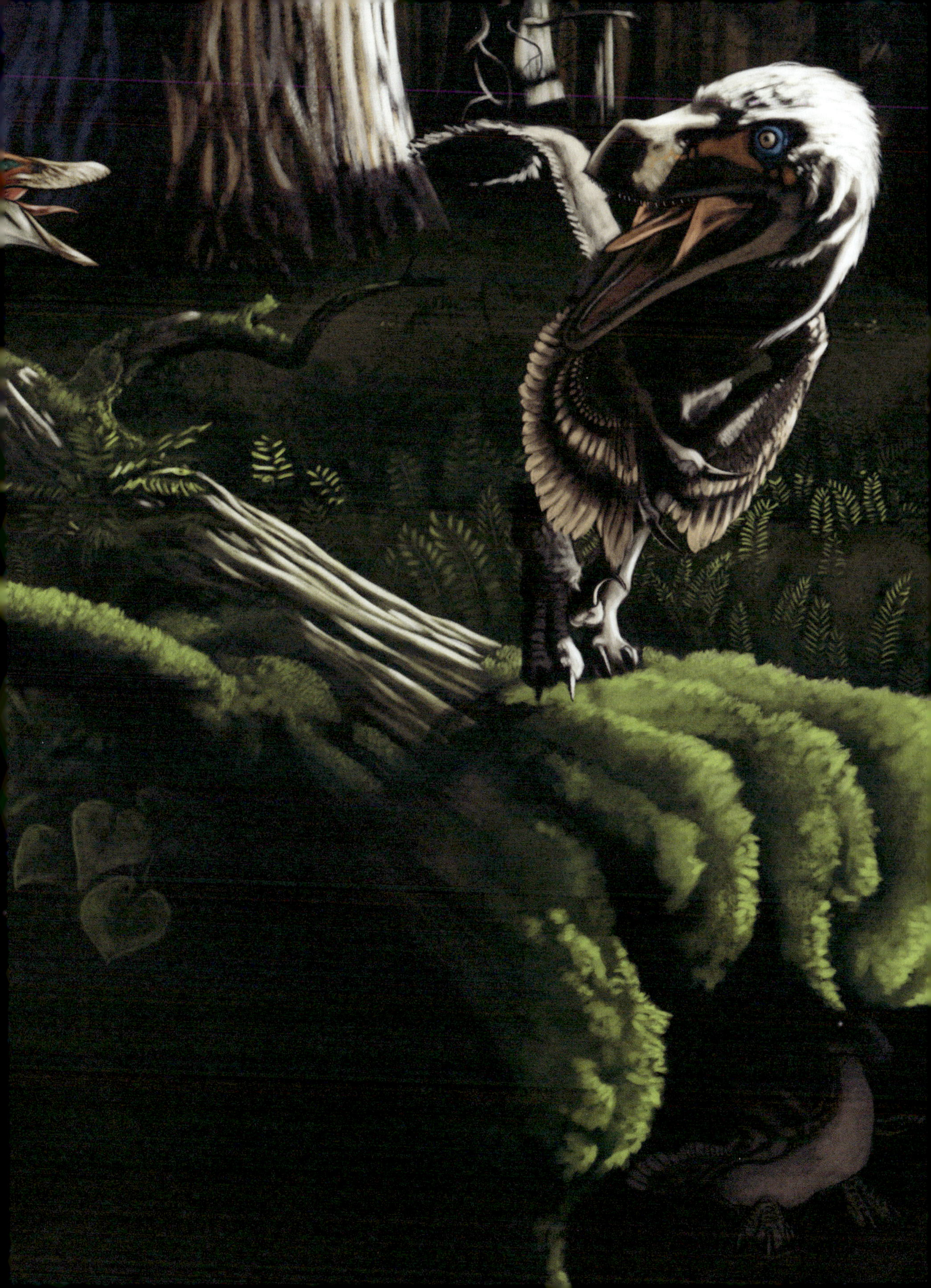

SELECT LIST OF SOURCES

Andrews, R.C. (2007). *Under A Lucky Star – A Lifetime Of Adventure*. Read Books.

Ashworth Jr, W.B. (1996). *Paper Dinosaurs 1824-1969, An Exhibition of Original Publications from the Collections of the Linda Hall Library*. Linda Hall Library.

Bakker, R.T. (1986). *The Dinosaur Heresies*. William Morrow and Company.

Bausum, A. (2000). *Dragon bones and dinosaur eggs: A photobiography of explorer Roy Chapman Andrews*. National Geographic Kids.

Benton, M.J. (2020). *The Dinosaurs Rediscovered, How a Scientific Revolution is Rewriting History*. Thames & Hudson.

Benton, M.J. (2021). *Dinosaurs: New Visions of a Lost World*. Thames & Hudson.

Bird, R.T. (1985). *Bones for Barnum Brown, Adventures of a Dinosaur Hunter*. Texas Christian University Press.

Bramwell, V. (2008). *All in the Bones: A Biography of Benjamin Waterhouse Hawkins*. Academy of Natural Sciences.

Brett-Surman, M.K., Holtz, T.R. & Farlow, J.O. (2012). *The Complete Dinosaur (Life of the Past)*. Indiana University Press.

Brusatte, S.L. (2012). *Dinosaur Paleobiology*. John Wiley & Sons.

Brusatte, S.L. (2018). *The Rise and Fall of the Dinosaurs*. Pan Macmillan.

Conway, J., Naish, D. & Kosemen, M. (2012). *All Yesterdays. Unique and Speculative Views of Dinosaurs and Other Prehistoric Animals*. Irregular Books.

Cordier, S. (2017). *De botten van de Borinage, de iguanodons van Bernissart van 125 miljoen voor Christus tot vandaag*. Vrijdag.

Dickens, C. (1850-1859). *Household Words*.

Dickens, C. (1852). Bleak House. Bradbury & Evans.

Doyle, A.C. (1912). *The Lost World*. Hodder & Stoughton.

Fastovsky, D.E., Weishampel, D.B. (2021). *Dinosaurs, A Concise Natural History*. Cambridge University Press.

Foster, J. (2007). *Jurassic West, Second Edition: The Dinosaurs of the Morrison Formation and Their World (Life of the Past)*. Indiana University Press.

Godefroit, P. (2012). *Bernissart Dinosaurs and Early Cretaceous Terrestrial Ecosystems*. Indiana University Press.

Holtz, T.R. & Rey, L.V. (2007). *Dinosaurs: The Most Complete, Up-to-Date Encyclopedia for Dinosaur Lovers of All Ages*. Random House.

Horner, J. & Gorman, J. (2009). *How to Build a Dinosaur: Extinction Doesn't Have to Be Forever*. Dutton Penguin.

Klein, N., Remes, K., Gee, C. & Martin, S. (2011). *Biology of the Sauropod Dinosaurs: Understanding the Life of Giants (Life of the Past)*. Indiana University Press.

Lescaze, Z. & Ford, W. (2017). *Paleoart. Visions of the Prehistoric Past*. Taschen.

Maier, G. (2003). *African Dinosaurs Unearthed: The Tendaguru Expeditions (Life of the Past)*. Indiana University Press.

Mantell, G. (1838). *The Wonders of Geology*. Henry G. Bohn.

McGowan, C. (1992). *Dinosaurs, Spitfires, and Sea Dragons*. Harvard University Press.

McGowan, C. (2002). *The Dragon Seekers. How an Extraordinary Circle of Fossilists Discovered the Dinosaurs and Paved the Way for Darwin*. Basic Books.

Mitchell, W.J.T. (1998). *The Last Dinosaur Book*. University of Chicago Press.

Naish, D. (2009). *The Great Dinosaur Discoveries*. A & C Black.

Naish, D. (2016). *Dinosaurs: How They Lived and Evolved*. Smithsonian Books.

Nothdurft, W. & Smith, J. (2002). *The Lost Dinosaurs of Egypt*. Random House.

Paul, G.S. (2016). *The Princeton Field Guide to Dinosaurs, Second Edition*. Princeton University Press.

Witton, M.P. (2018). *Recreating an Age of Reptiles*. Crowood Press.

Witton, M.P. (2019). *The Paleoartist's Handbook. Recreating prehistoric animals in art*. Crowood Press.

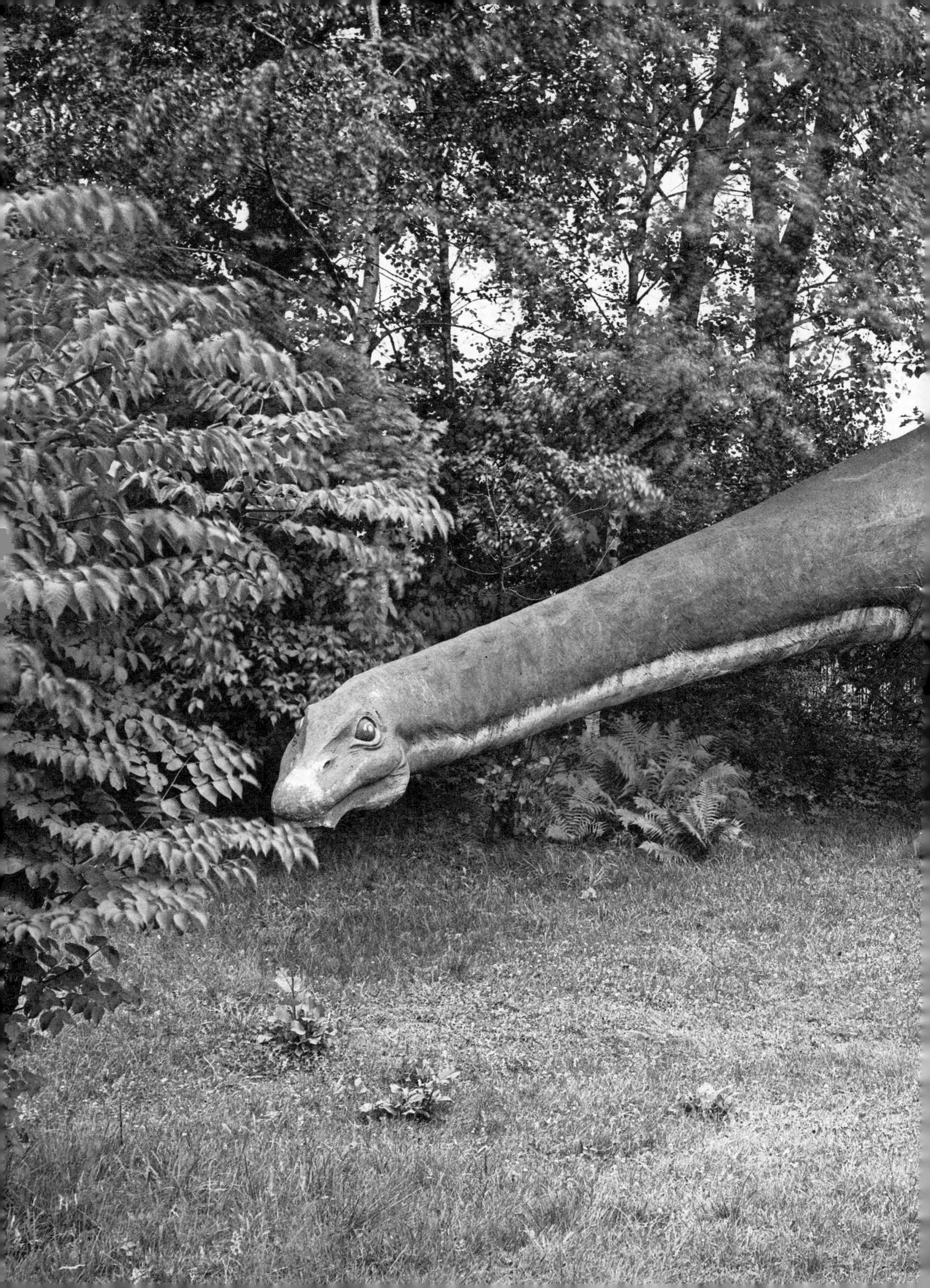

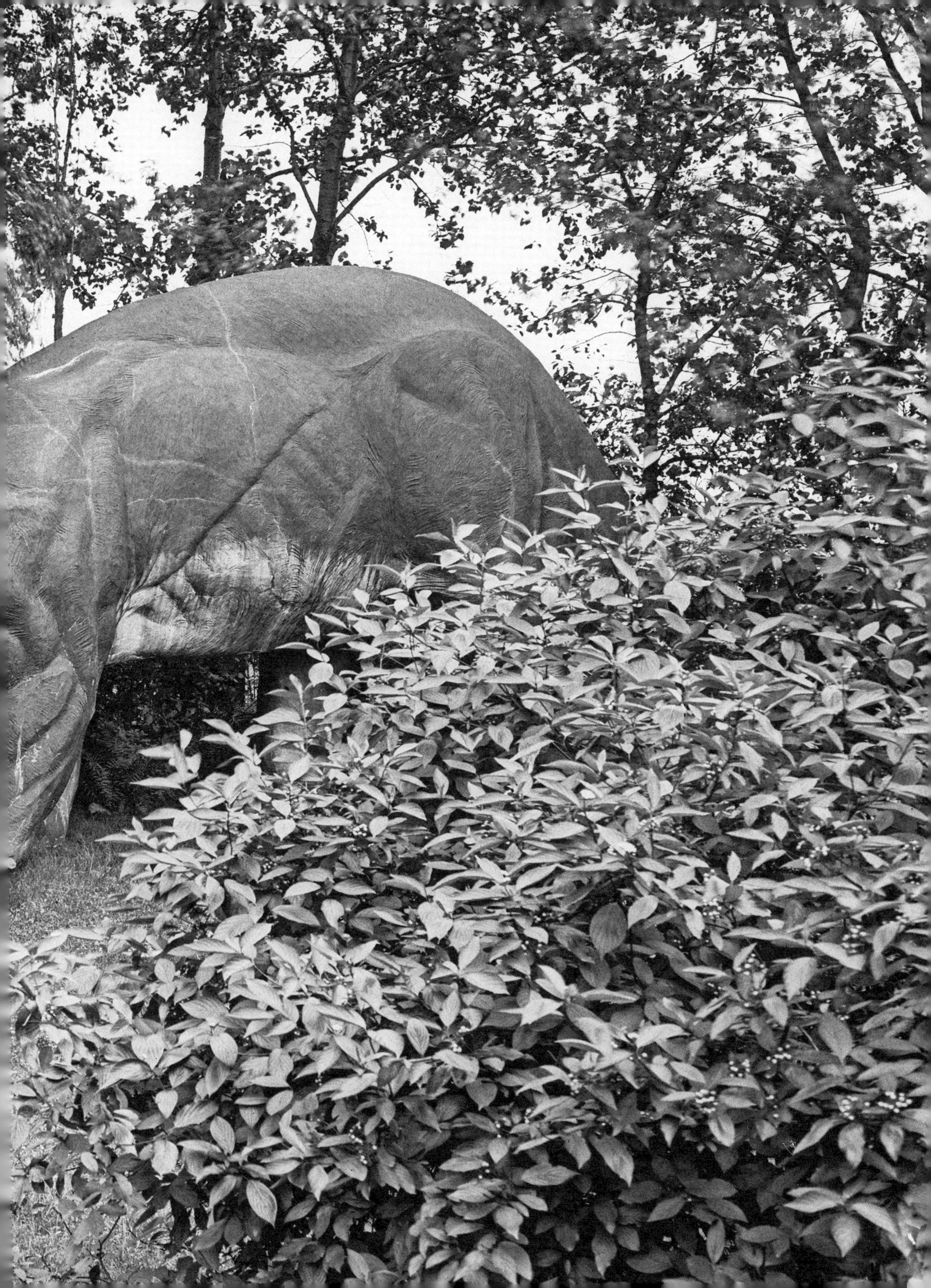

<
A replica of the *Diplodocus carnegiei* in the Tierpark Hagenbeck in Hamburg (1909).

>
Triceratops prorsus, the last of the dinosaurs', a lithograph after an illustration
from 'Extinct Monsters and Creatures of Other Days' (1894)

>>
Ichtyosaurus, Tom Liekens (2018).

CREDITS

Cover: Illustration by J. Smit from H. N. Hutchinson's *Extinct Monsters and Creatures of Other Days*, Chapman and Hall, London, 1894, © Florilegius / Alamy / Imageselect; p. 2 Imageselect; pp. 4 – 5 © Tom Liekens; pp. 6 – 7 © Juri de Luca & Gabriele Galimberti; p. 8 © Arthur Pollock; p. 11 © Zoic; pp. 12 – 13 © NHM/Collection Thomas Hofmann; pp. 16 – 17 © Christie's Images / Bridgeman Images; p. 20 photography: Vincent Girier-Dufournier for Binoche et Giquello; p. 21 photography: Vincent Girier-Dufournier for Binoche et Giquello; pp. 22 – 23 © Teatrum Mundi Gallery, picture by Juri De Luca and Gabriele Galimberti for National Geographic 2019; p. 25 © Laugery @ Les Jardins de Marqueyssac-Dordogne; p. 27 © Granada Gallery; p. 29 © Imageselect; p. 30 © Agence Rol: agence photographique - 1908 - National Library of France, France; p. 33 © Imageselect; pp. 36 – 37 © Florilegius / Bridgeman Images; p. 38 Imageselect; pp. 42 – 43 © Science Photo Library; p. 45 © Imageselect; p. 46 © Florilegius / Alamy / Imageselect; p. 48 © The Natural History Museum / Alamy / Imageselect; p. 49 © The Picture Art Collection / Alamy / Imageselect; pp. 50 – 51 © Wellcome Collection; p. 53 © World History Archive / Imageselect; p. 54 © The Natural History Museum / Alamy / Imageselect; p. 57 Photograph by Maull & Polyblank. © Wellcome Collection; p. 59 Wiki Commons; pp. 60 – 61 USNM V 2580, Image Courtesy of the Smithsonian Institution; p. 63 (left) © Alpha Historica / Alamy / Imageselect, (right) © GL Archive / Alamy / Imageselect; p. 65 © Royal Institute of Natural Sciences, Brussels; pp. 66 – 67 © Royal Institute of Natural Sciences, Brussels; p. 69 ©Royal Institute of Natural Sciences, Brussels ; p. 71 © American Museum of Natural History; p. 73 © FLHC DC1 / Alamy / Imageselect; p. 75 Science Photo Library; p. 76 © Florilegius / Alamy / Imageselect; pp. 80 – 81 © Science Photo Library; pp. 84 – 85 © Tom Liekens; pp. 88 – 89 © Zoic; © F.L.C. – SABAM Belgium 2022; pp. 94 – 95 © GL Archive / Alamy / Imageselect; p. 96 © Princeton University, Department of Geosciences, Guyot Hall; pp. 98 – 99 © The Picture Art Collection / Alamy / Image Select; pp. 102 – 103 © Andrey Atuchin; pp. 106 – 107 courtesy of the artist Kenny Scharf, photo: Joshua White photography, © SABAM Belgium 2022; pp. 108 – 109 Archive Galerie Max Hetzler Berlin | Paris | London © Albert Oehlen, SABAM Belgium 2022; p. 110 © Tom Liekens; pp. 112 – 113 © Tom Liekens; p. 114 Courtesy of the Artists, Jake and Dinos Chapman, © SABAM Belgium 2022; p. 115 Courtesy of the Artists, Jake and Dinos Chapman, © SABAM Belgium 2022; pp. 116 – 117 Courtesy of the artist and Tanya Bonakdar; p. 119 photo: Jeff Mclane, Courtesy of the Artist and The Pit; p. 121 © Panamarenko Foundation, SABAM Belgium 2022; p. 122 Courtesy Gallery Sofie Van de Velde; p. 123 Courtesy Gallery Sofie Van de Velde; pp. 124 – 125 Courtesy Gallery Sofie Van de Velde; pp. 126 – 127 © Imageselect; p. 128 © Matt Johnson, Courtesy of the artist and Blum & Poe, Los Angeles/New York/Tokyo; p. 129 © Darwin, Sinke & van Tongeren; pp. 130 – 131 © Darwin, Sinke & van Tongeren; p. 133 © Imageselect; p. 134 © Imageselect; pp. 138 – 139 © Imageselect; p. 140 © ANSP Archives Collection 803; p. 143 © Imageselect; p. 144 – 145 © Imageselect; p. 147 Wiki Commons; pp. 148 – 149 © Science Photo Library; pp. 152 – 153 © Royal Institute of Natural Sciences, Brussels; p. 155 © Imageselect; pp. 156 – 157 © Juri de Luca & Gabriele Galimberti; p. 161 Wiki Commons; p. 162 © Ward Broes; pp. 163 – 164 © Ward Broes; pp. 165 – 166 © SLUB Dresden / Deutsche Fotothek / Paul Schulz; pp. 169 – 170 © Florilegius / Alamy / Imageselect; pp. 173 – 174 © Tom Liekens

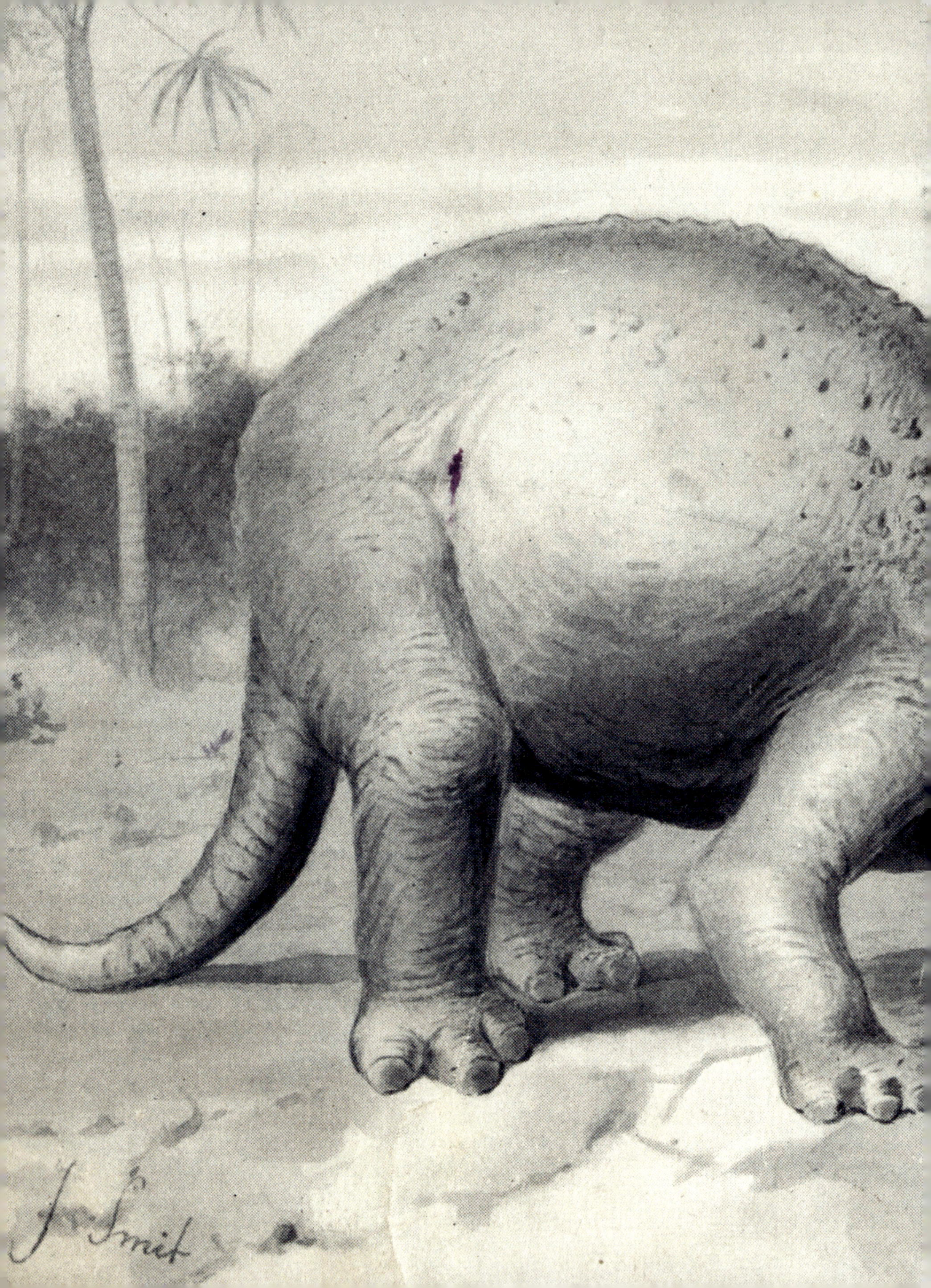

Jos Smit

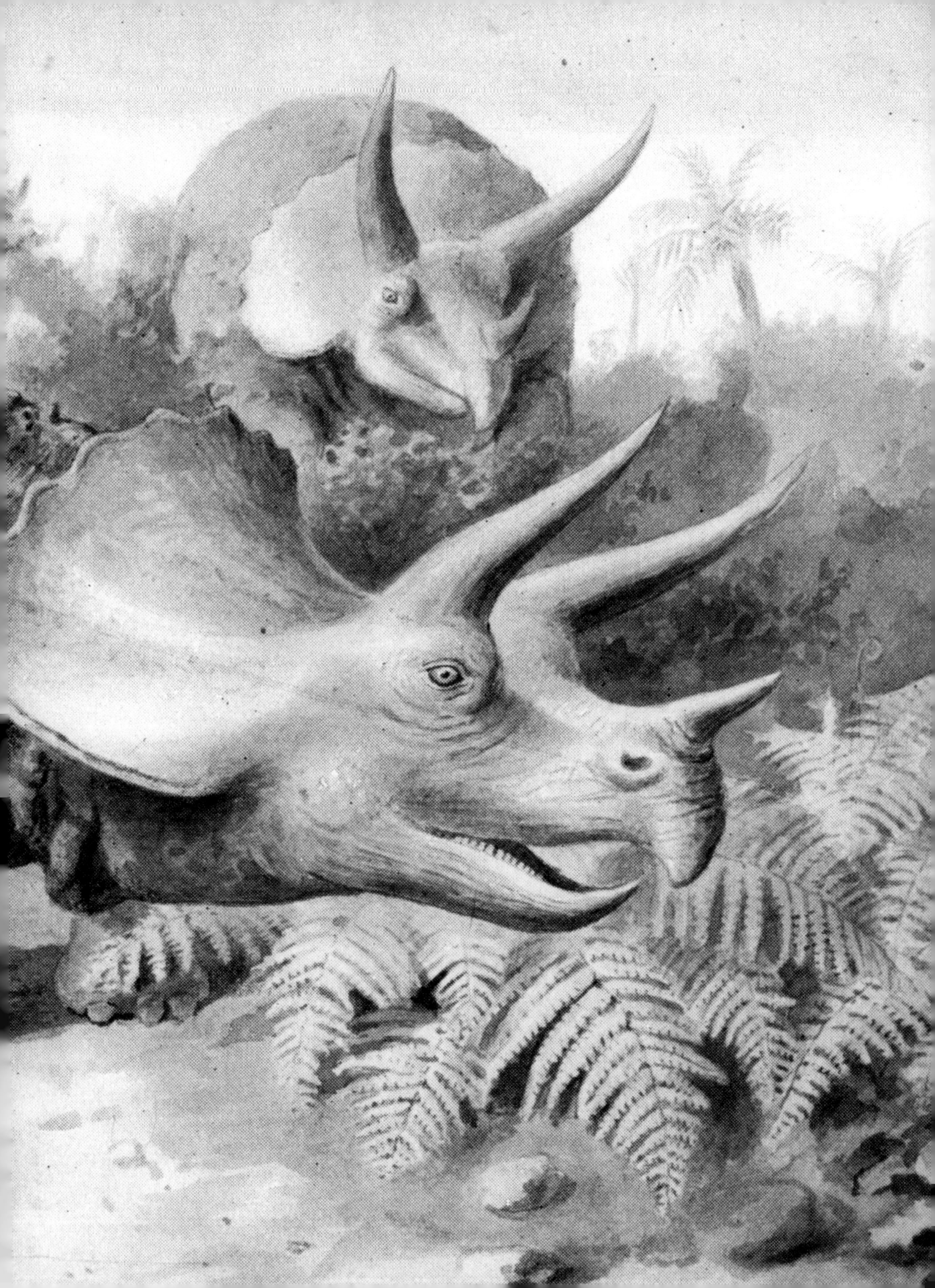

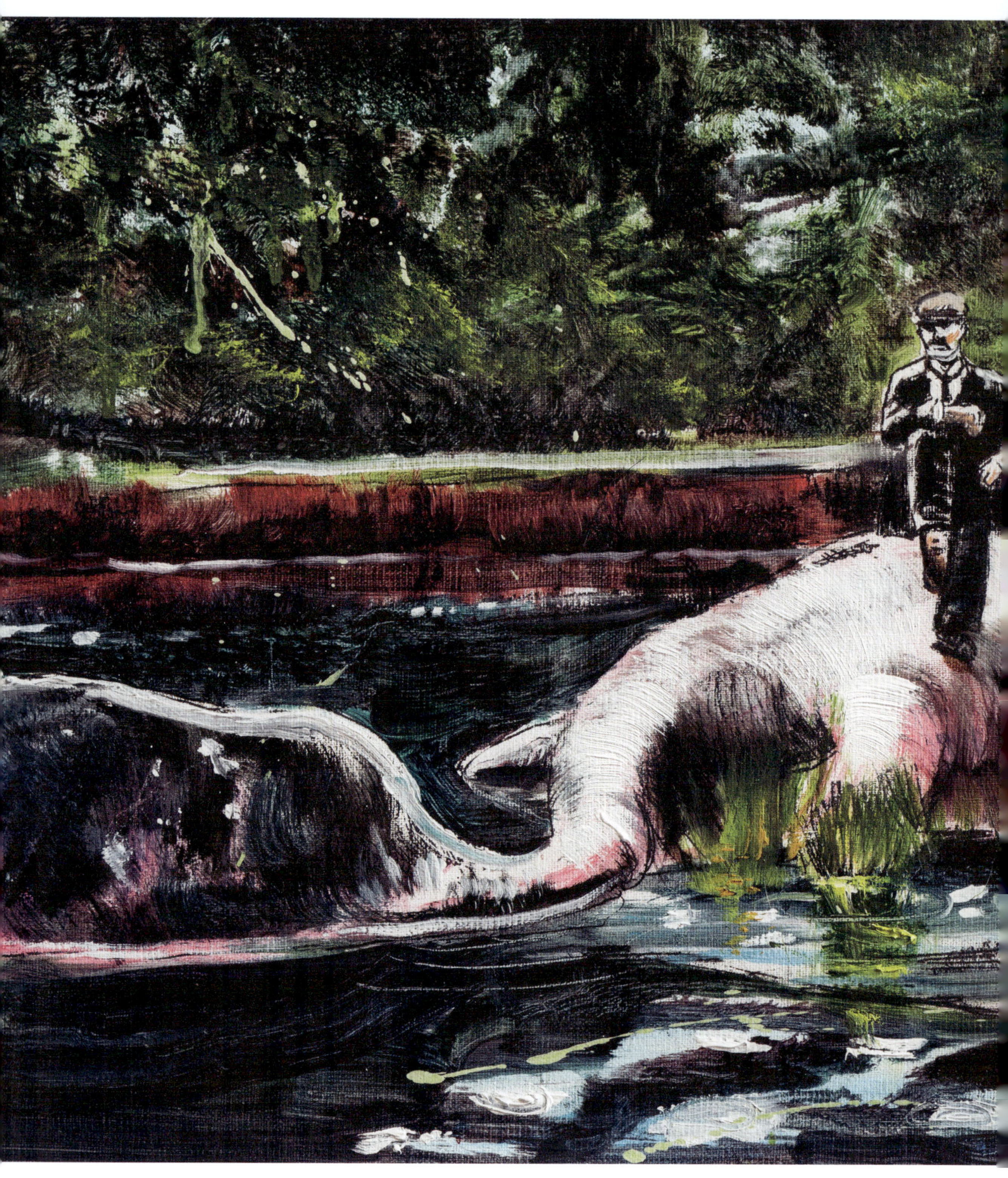

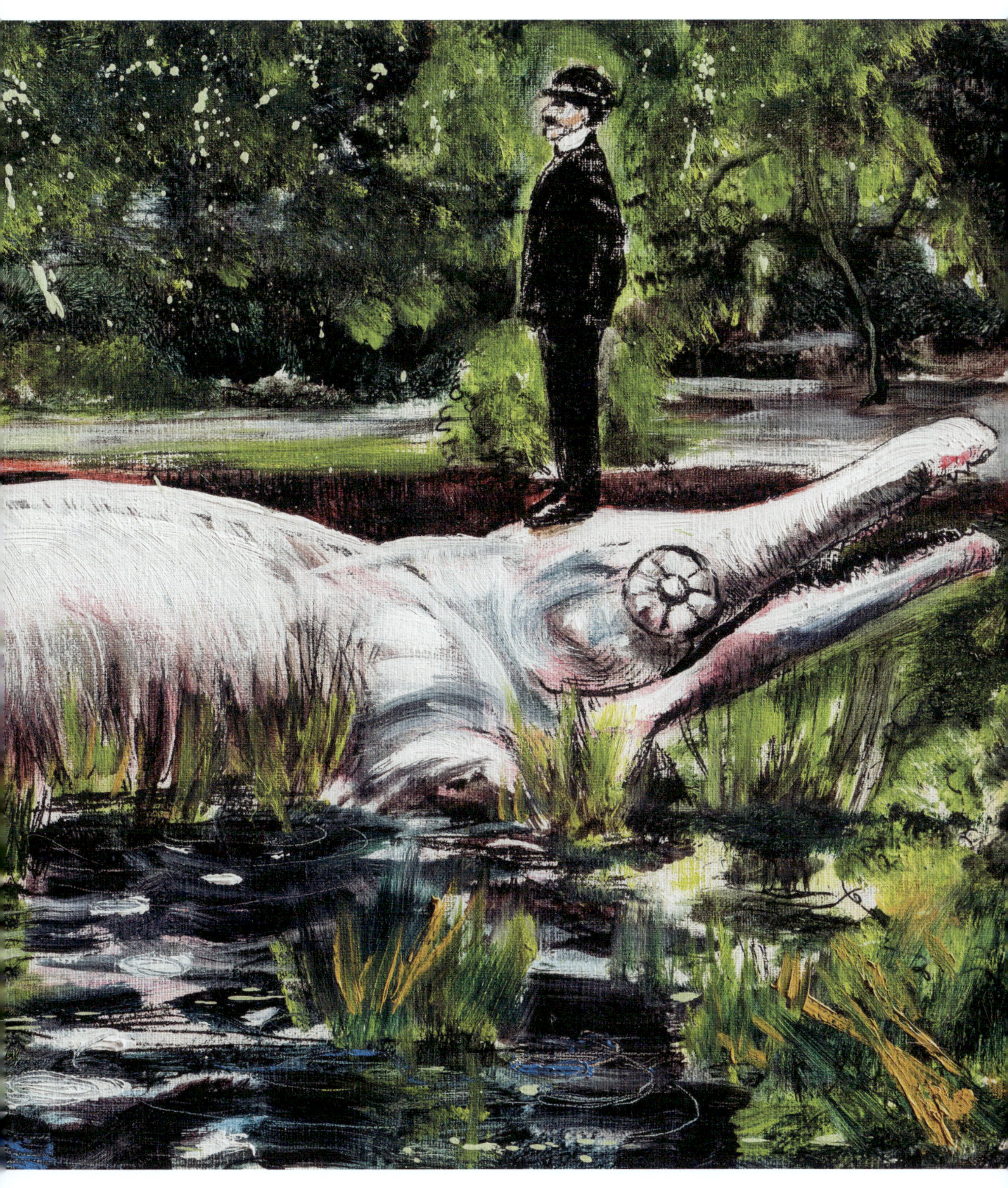

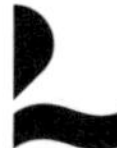

WWW.LANNOO.COM
Sign up to our newsletter for updates on
our latest publications on art, interior design,
food & travel, photography and fashion,
as well as exclusive offers and events.

TEXTS
Thijs Demeulemeester
Koen Stein

TRANSLATION
Lisa Holden

COPY EDITING
Léa Teuscher

IMAGE SELECTION
Laurence Vander Elstraeten (Buro Bonito)
Sarah Theerlynck

BOOK DESIGN
Bart Luijten

If you have any questions or comments about the material in this book,
please do not hesitate to contact our editorial team: art@lannoo.com

© Lannoo Publishers, Belgium, 2022
D/2022/45/23 - NUR: 640/682
ISBN 9789401482158

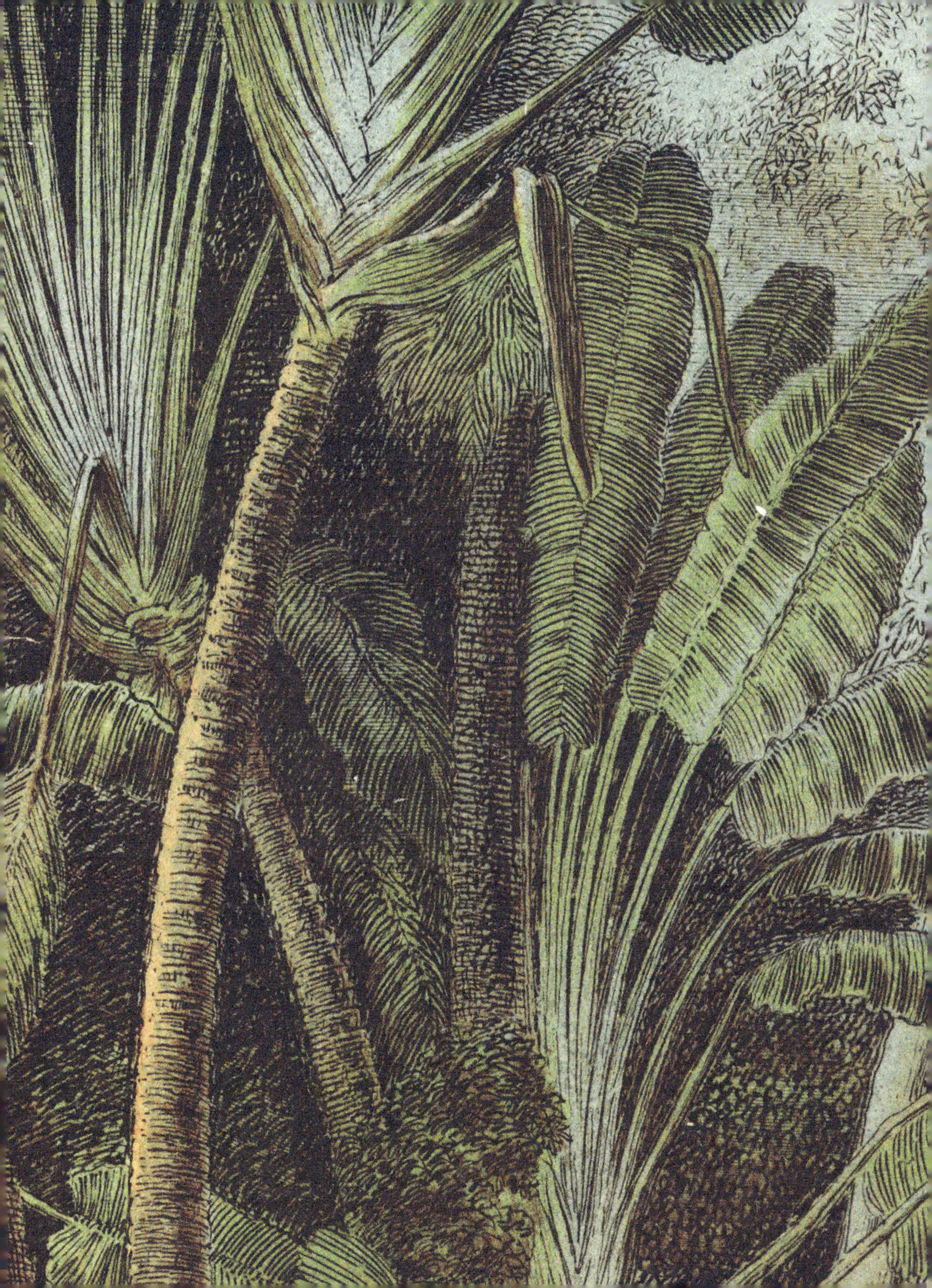

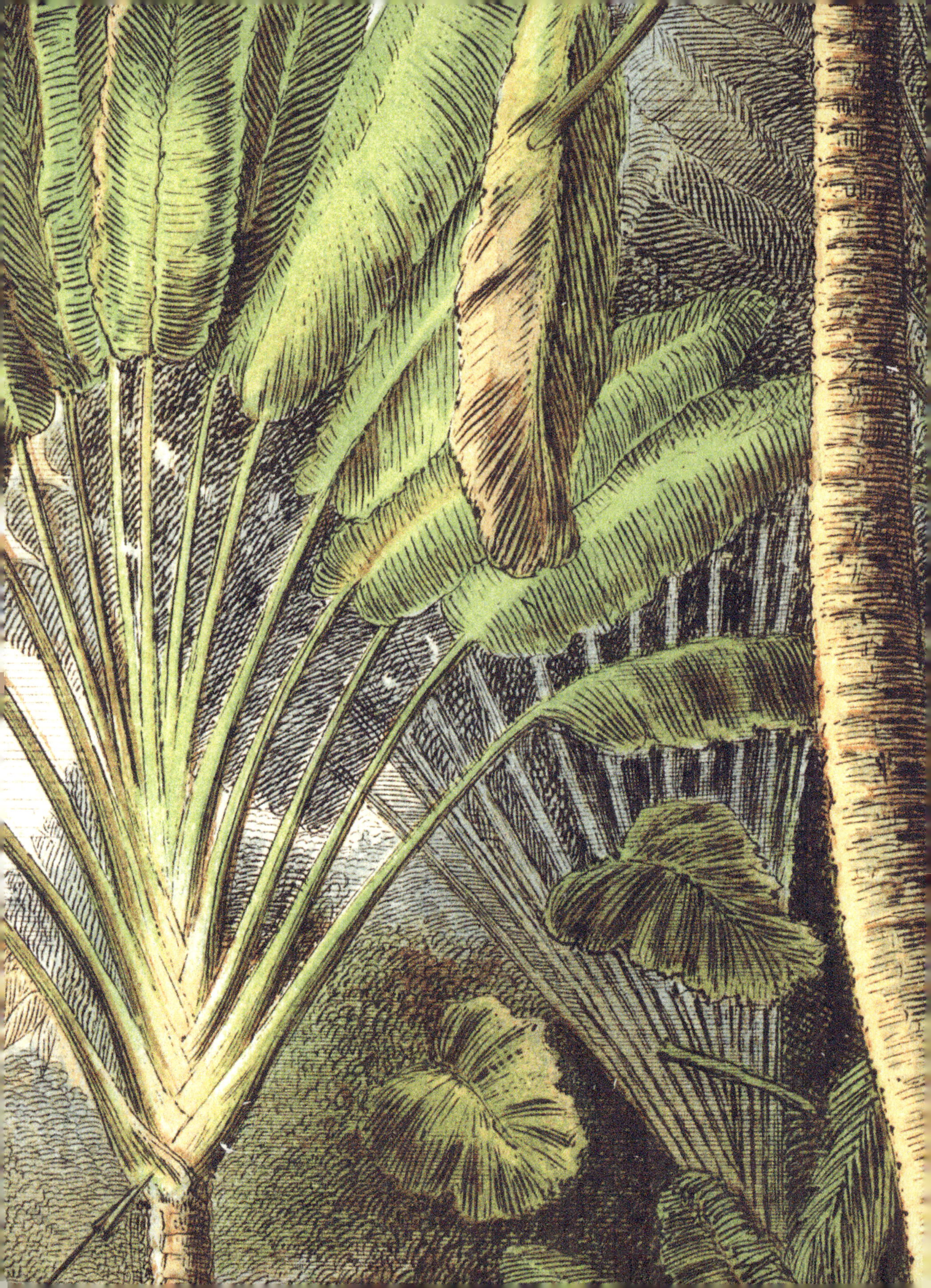